AS Level Maths is Really Hard

AS level maths is seriously tricky — no question about that.

We've done everything we can to make things easier for you. We've scrutinised past paper questions and we've gone through the syllabuses with a fine-toothed comb. So we've found out exactly what you need to know, then explained it simply and clearly.

We've stuck in as many helpful hints as you could possibly want — then we even tried to put some funny bits in to keep you awake.

We've done our bit — the rest is up to you.

What CGP's All About

The central aim of Coordination Group Publications is to produce top quality books that are carefully written, immaculately presented and astonishingly witty — whilst always making sure they exactly cover the syllabus for each subject.

And then we supply them to as many people as we possibly can, as <u>cheaply</u> as we possibly can.

Contents

Section One — Kinematics

Section Two — Vectors

Section Three — Statics

Section Four — Dynamics

Section Five — Projectiles

NB — In this book, if it's not specified, take the force of gravity to be 9.8 ms^{-2} ($g = 9.8\ ms^{-2}$).

This book covers all the major topics for Edexcel, OCR A, OCR MEI, AQA A and AQA B boards. There are notes at the top of some pages to tell you if there's a bit you can ignore.

Published by Coordination Group Publications Ltd
Typesetters:
Martin Chester, Sharon Watson
Contributors:
Andy Ballard, Janet Dickinson, Mark Moody, Rob Savage
Editors:
Charley Darbishire, Simon Little, Tim Major and Andy Park
Many thanks to Sharon Keeley
and Janet Dickinson *for proofreading.*

ISBN 1 84146 986 6
Groovy website: www.cgpbooks.co.uk
Jolly bits of clipart from CorelDRAW
Printed by Elanders Hindson, Newcastle upon Tyne.

Constant Acceleration Equations

Welcome to the technicolour world of Mechanics 1. Fashions may change, but there will always be M1 questions that involve objects travelling in a straight line. It's just a case of picking the right equations to solve the problem.

There are Five Constant Acceleration Equations

Examiners call these "uvast" questions (pronounced ewe-vast, like a large sheep) because of the five variables involved:

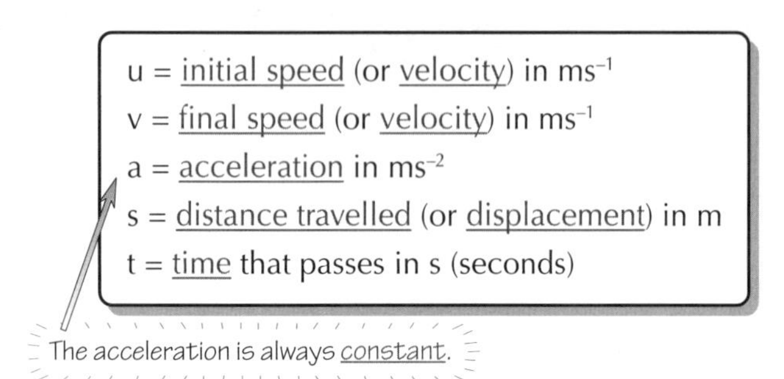

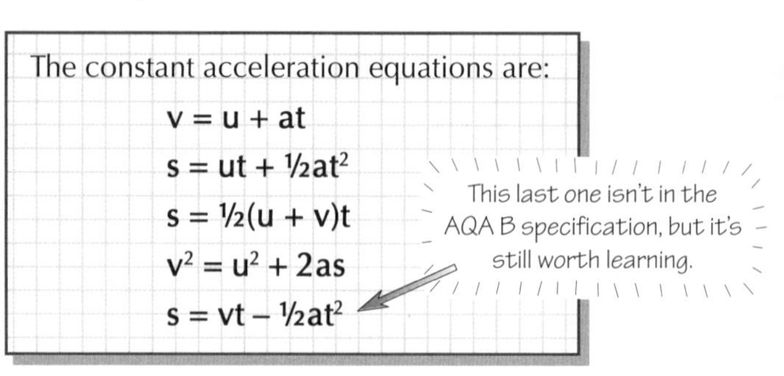

None of those equations are in the formula book, so you're going to have to learn them. Questions will usually give you three variables — your job is to choose the equation that will find you the missing fourth variable.

Example: A jet ski travels in a straight line along a river. It passes under two bridges 200 m apart and is observed to be travelling at 5 ms^{-1} under the first bridge and at 9 ms^{-1} under the second bridge. Calculate its acceleration (assuming it is constant).

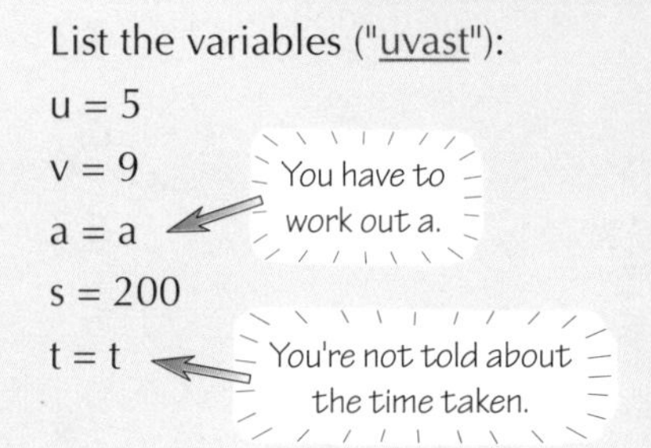

Check you're using the right units — m, s, ms^{-1} and ms^{-2}.

Choose the equation with u, v, s and a in it: $v^2 = u^2 + 2as$

Substitute values: $9^2 = 5^2 + (2 \times a \times 200)$

Simplify: $81 = 25 + 400a$

Rearrange: $400a = 81 - 25 = 56$

Then solve: $a = \frac{56}{400} = 0.14 \text{ ms}^{-2}$

Motion under Gravity just means taking a = g

Use the value of g given on the front of the paper or in the question. If you don't, you risk losing a mark because your answer won't match the examiners' answer.

Don't be put off by questions involving objects moving freely under gravity — they're just telling you the acceleration is g.

Example: A pebble is dropped into a well 18 m deep and moves freely under gravity until it hits the bottom. Calculate the time it takes to reach the bottom. (Take g = 9.8 ms^{-2}.)

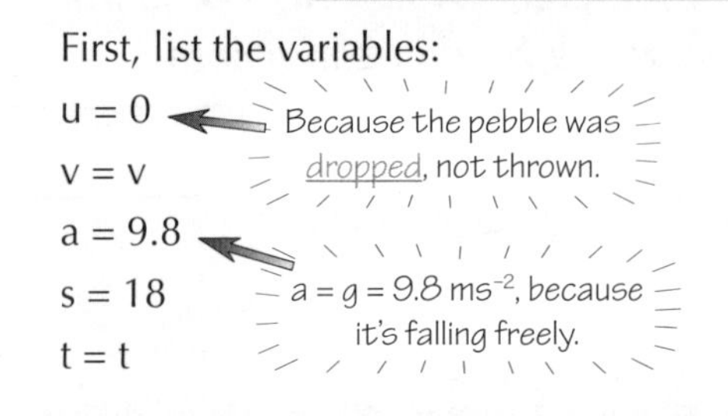

You need the equation with u, a, s and t in it: $s = ut + \frac{1}{2}at^2$

Substitute values: $18 = (0 \times t) + (\frac{1}{2} \times 9.8 \times t^2)$

Simplify: $18 = 4.9t^2$

Rearrange to give t^2: $t^2 = \frac{18}{4.9} = 3.67...$

Solve by square-rooting: $t = \sqrt{3.67...} = 1.92 \text{ s}$

Watch out for tricky questions like this — at first it looks like they've only given you one variable.
You have to spot that the pebble was dropped (so it started with no velocity) and that it's moving freely under gravity.

Constant Acceleration Equations

Sometimes there's More Than One Object Moving at the Same Time

For these questions, t is often the same (or connected as in this example) because time ticks along for both objects at the same rate. The distance travelled might also be connected.

Example:

A car, A, travelling along a straight road at a steady 30 ms^{-1} passes point R at t = 0. Exactly 2 seconds later, a second car, B, travelling at 25 ms^{-1}, moves in the same direction from point R. Car B accelerates at a constant 2 ms^{-2}. Show that the two cars are level when $\mathbf{t^2 - 9t - 46 = 0}$ where t is the time taken by car A.

For each car, there are different "uvast" equations, so you write separate lists and separate equations.

CAR A	CAR B
$u_A = 30$	$u_B = 25$
$v_A = 30$	$v_B = v$
$a_A = 0$	$a_B = 2$
$s_A = s$	$s_B = s$
$t_A = t$	$t_B = (t - 2)$

Constant speed so $a_A = 0$

s is the same for both cars because they're level.

B starts moving 2 seconds after A passes point R.

The two cars are level, so choose an equation with s in it:

$s = ut + \frac{1}{2}at^2$

Substitute values: $s = 30t + (\frac{1}{2} \times 0 \times t^2)$

Simplify: $\mathbf{s = 30t}$

Use the same equation for Car B: $s = ut + \frac{1}{2}at^2$

Substitute values: $s = 25(t - 2) + (\frac{1}{2} \times 2 \times (t - 2)^2)$

Simplify: $s = 25t - 50 + (t - 2)(t - 2)$

$s = 25t - 50 + (t^2 - 4t + 4)$

$\mathbf{s = t^2 + 21t - 46}$

The distance travelled by both cars is equal, so put the equations for s equal to each other:

$30t = t^2 + 21t - 46$

$t^2 - 9t - 46 = 0$

That's the result you were asked to find.

Constant acceleration equation questions involve modelling assumptions (simplifications to real life so you can use the equations):

1) The object is a particle — this just means it's very small and so isn't affected by air resistance as cars or stones would be in real life.

2) Acceleration is constant — without it, the equations couldn't be used.

Practice Questions

1) A motorcyclist accelerates uniformly from 3 ms^{-1} to 9 ms^{-1} in 2 seconds. What is the distance travelled by the motorcyclist during this acceleration?

2) A ball is projected vertically upwards at 3 ms^{-1}. How long does it take to reach its maximum height?

3) Sample exam question:

The window cleaner of a high-rise block accidentally drops his sandwich, which then falls freely to the ground. The height between the consecutive floors of the building is h and the speed of the sandwich as it passes a high floor is u. The sandwich takes 1.2 seconds to fall a further 4 floors. Use g = 10ms^{-2}.

a) Show that $4h = 1.2u + 7.2$ [2 marks]

b) The sandwich takes another 0.6 seconds to fall a further 4 floors.

i) Obtain another equation in u and h. [2 marks]

ii) Hence calculate the values of u and h. [2 marks]

c) Comment on modelling assumptions you have made. [1 mark]

As Socrates once said, "The unexamined life is not worth living"... *but what did he know...*

Make sure you:

1) Make a list of the uvast variables EVERY time you get one of these questions.

2) Look out for "hidden" values — e.g. "particle initially at rest..." means u = 0.

3) Choose and solve the equation that goes with the variables you've got.

Motion Graphs

You can use displacement-time (t, x), velocity-time (t, v) and acceleration-time (t, a) graphs to represent all sorts of motion.

Displacement-time Graphs — Height = Distance and Gradient = Velocity

The steeper the line, the greater the velocity. A horizontal line has a zero gradient, so that means the object isn't moving.

Example: A cyclist's journey is shown on this (t,x) graph. Describe the motion.

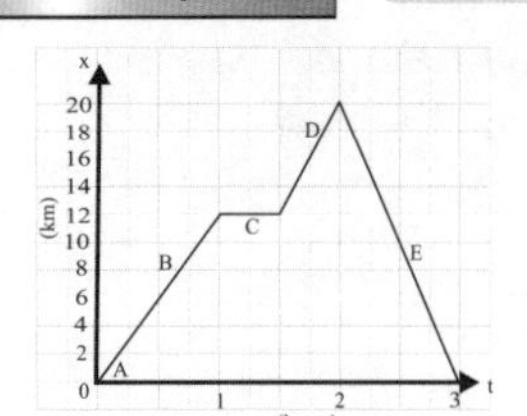

A: Starts from rest (when t = 0, x = 0)

B: Travels 12 km in 1 hour at a velocity of 12 kmh^{-1}

C: Rests for ½ hour (v = 0)

D: Cycles 8 km in ½ hour at a velocity of 16 kmh^{-1}

E: Returns to starting position, cycling 20 km in 1 hour at a velocity of -20 kmh^{-1} (i.e. 20 kmh^{-1} in the opposite direction)

Example: A girl jogs 2 km in 15 minutes and a boy sprints 1.5 km in 6 min, rests for 1 min then walks the last 0.5 km in 8 min. Show the two journeys on a (t,x) graph.

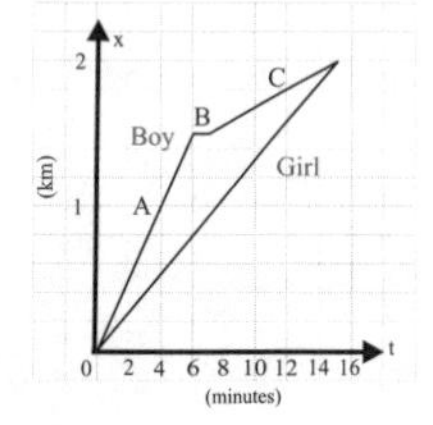

Girl: constant velocity, so there's just one straight line for her journey from (0,0) to (15,2)
Boy: three parts to the journey, so there's three straight lines: A - sprint, B - rest, C - walk

Velocity-time Graphs — Area = Distance and Gradient = Acceleration

The area under the graph can be calculated by splitting the area into rectangles, triangles or trapeziums. Work out the areas separately, then add them all up at the end.

Example: A train journey is shown on the (t,v) graph on the right. Find the distance travelled and the rate of deceleration as the train comes to a stop.

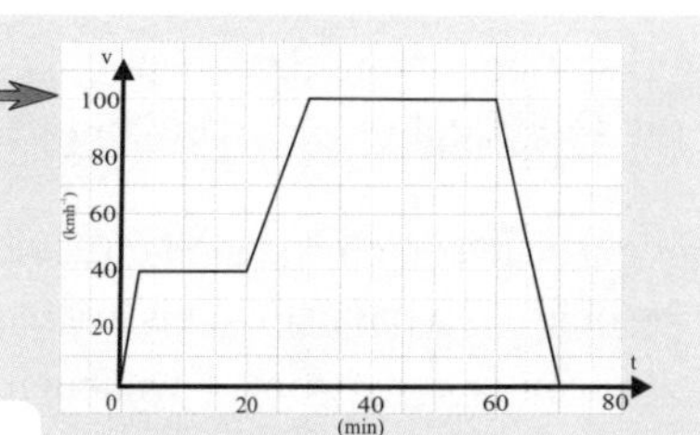

The time is given in minutes and the velocity as kilometres per hour, so divide the time in minutes by 60 to get the time in hours.

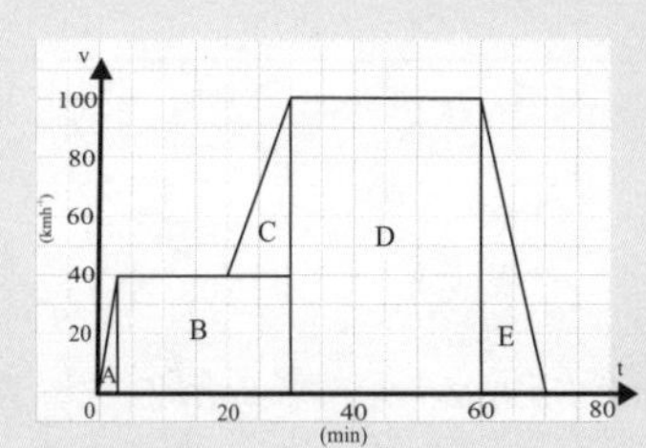

Area of A: $(2.5 \div 60 \times 40) \div 2 = 0.83$

Area of B: $27.5 \div 60 \times 40 = 18.33$

Area of C: $(10 \div 60 \times 60) \div 2 = 5$

Area of D: $30 \div 60 \times 100 = 50$

Area of E: $(10 \div 60 \times 100) \div 2 = 8.33$

Total area = 82.5 so distance is 82.5 km

The gradient of the graph at the end of the journey is $-100\ kmh^{-1} \div (10 \div 60)$ hours $= -600\ kmh^{-2}$
So the train decelerates at 600 kmh^{-2}.

Acceleration-time Graphs — Area = Velocity

Example: The acceleration of a parachutist who jumps from a plane is shown on the (t,a) graph on the right. Describe the motion of the parachutist and find the velocity as the parachutist falls towards the ground.

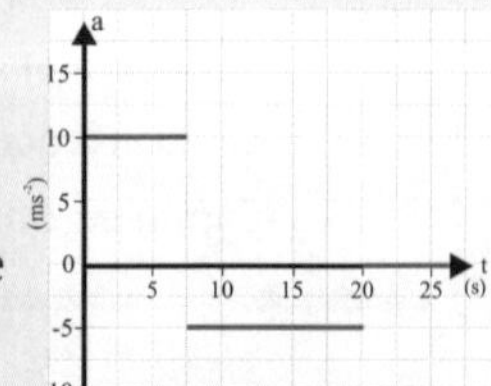

He falls with acceleration due to gravity of 10 ms^{-2} for 7.5 s. The parachute opens and the acceleration due the air resistance of the parachute is 5 ms^{-2} acting upwards for 12.5 s. After 20 s, the acceleration is zero and so he falls with constant velocity. You need to find the area under the graph:

Area A: $10 \times 7.5 = 75\ ms^{-1}$ Area B: $5 \times 12.5 = 62.5\ ms^{-1}$

Area B is under the horizontal axis, so subtract area B from area A:

Velocity = $75\ ms^{-1} - 62.5\ ms^{-1} = 12.5\ ms^{-1}$

Motion Graphs

Graphs can be used to Solve Complicated Problems

As well as working out distance, velocity and acceleration from graphs, you can also solve more complicated problems. These might involve working out information not shown directly on the graph.

Example: A jogger and a cyclist set off at the same time. The jogger runs with a constant velocity. The cyclist accelerates from rest, reaching a velocity of 5 ms^{-1} after 6 s and then continues at this velocity. The cyclist overtakes the jogger after 15 s.

Draw a graph of the motion and find the velocity of the jogger.

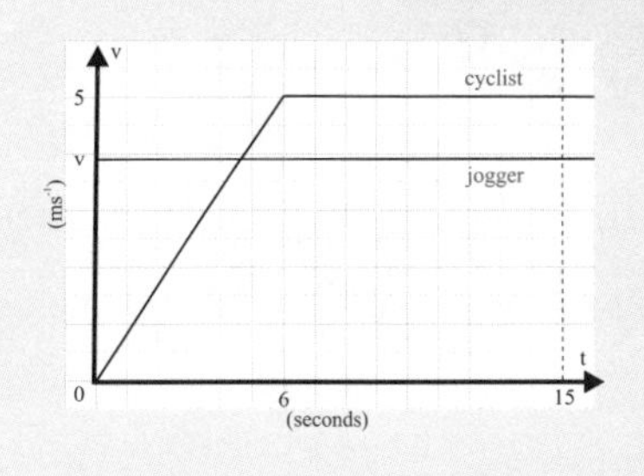

Call the velocity of the jogger v.

After 15 s the distance each has travelled is the same, so you can work out the area under the two graphs to get the distances:

Jogger: Area = distance = 15v

Cyclist: Area = distance = $(5 \times 6) \div 2 + (9 \times 5) = 60$

(area of triangle + area of rectangle)

So $15v = 60$

$v = 4\ ms^{-1}$

Practice Questions

1) A runner starts from rest and accelerates at 0.5 ms^{-2} for 5 seconds. She maintains a constant velocity for 20 seconds then decelerates to a stop at 0.25 ms^{-2}.

Draw a (t,v) graph to show the motion and find the distance the runner travelled.

2) The start of a journey is shown on the (t,a) graph below.

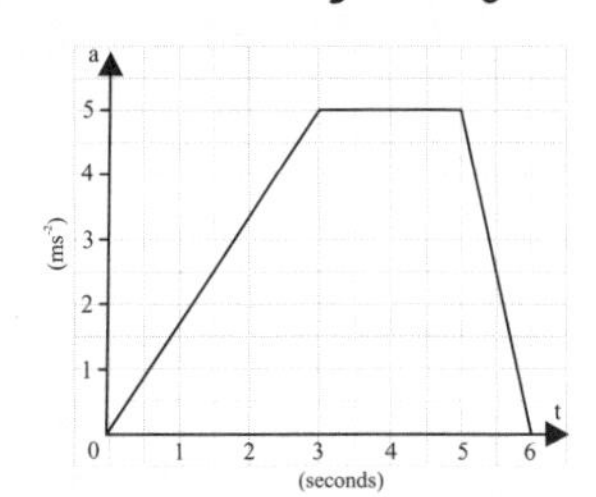

Find the velocity when: a) t = 3 b) t = 5 c) t = 6

3) Sample exam question:

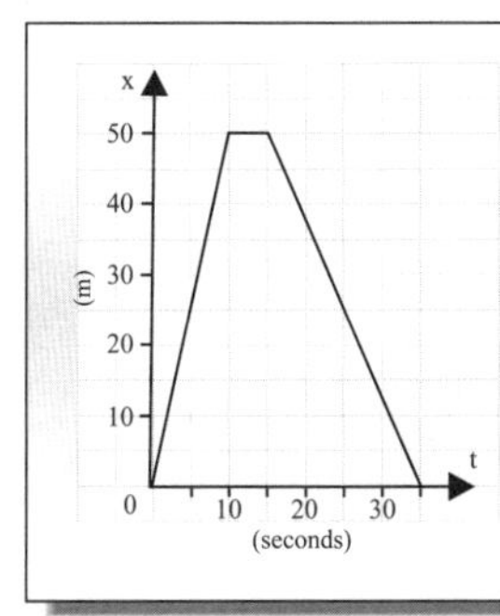

A sprinter runs 50 m, rests and then jogs back. His motion is shown on this (x,t) graph.

a) When is the runner at rest? Explain how the graph shows this. [1 mark]

b) Find the runner's velocity in the first part of the motion. [2 marks]

c) What is the runner's velocity as he returns home? [2 marks]

Random tongue-twister #1 — I wish to wash my Irish wristwatch...

If a picture can tell a thousand words then a graph can tell... um... a thousand and one. Make sure you know what type of graph you're using, and learn what the gradient and the area under each type of graph tell you.

Variable Acceleration

Skip these two pages if you're doing Edexcel or AQA B.

Find out all there is to know about how something is moving — use differentiation and integration to switch between position, velocity and acceleration.

Differentiating lets you change from Position to Velocity to Acceleration

The position, velocity and acceleration of a moving object can be written as expressions in terms of time, t. Velocity is the rate of change of position, so you can differentiate position to find velocity.

In the same way, acceleration is the rate of change of velocity, so differentiating velocity gives you acceleration.

Example: An object moves in a straight line. Its position at time t is given by $r = 3t^2 + 5t - 1$. Find the velocity and acceleration of the object at time t.

$$\frac{d}{dt}(3t^2+5t-1) = 6t+5 \quad \text{so} \quad v = 6t + 5$$

$$\frac{d}{dt}(6t+5) = 6 \quad \text{so} \quad a = 6$$

Differentiate Both Components of a Vector Separately

See Section 2 for more about vectors.

Example: The position vector of an object is $\boldsymbol{r} = \begin{bmatrix} 7t+3 \\ 5t^2-2t+1 \end{bmatrix}$. Find the velocity and acceleration of the object when t = 0.

The top line is movement parallel to the x-axis, the bottom line parallel to the y-axis.

Velocity: $\frac{d}{dt}(7t+3) = 7$ and $\frac{d}{dt}(5t^2-2t+1) = 10t - 2$ so $\boldsymbol{v} = \begin{bmatrix} 7 \\ 10t-2 \end{bmatrix}$

when t = 0, $\boldsymbol{v} = \begin{bmatrix} 7 \\ 10\times0-2 \end{bmatrix} = \begin{bmatrix} 7 \\ -2 \end{bmatrix}$

Acceleration: $\frac{d}{dt}7 = 0$ and $\frac{d}{dt}(10t-2) = 10$ so $\boldsymbol{a} = \begin{bmatrix} 0 \\ 10 \end{bmatrix}$

When t = 0, $\boldsymbol{a} = \begin{bmatrix} 0 \\ 10 \end{bmatrix}$ i.e. the acceleration is constant (the same no matter what value t is).

Work Backwards using Integration

I'm sure this is more than a little obvious to you, but integration is the inverse of differentiation, so you can integrate acceleration to get velocity, and integrate velocity to get position.

Example: An object moves in a straight line with velocity in ms^{-1} at time t given by $v = 3t^2 + 5$. Find the distance (r) it travels from t = 3 to t = 5.

$$r = \int_{t=3}^{t=5} v\,dt = \int_{t=3}^{t=5} (3t^2+5)\,dt$$

Use the values of t as limits.

$$= \left[t^3 + 5t\right]_{t=3}^{t=5}$$

= (125 + 25) – (27 + 15)

= 108 m

Variable Acceleration

Use Values to work out General Formulas when Integrating

If you integrate to find a general formula (i.e. when there are no limits), you have to add on a constant. If you know extra information such as the initial velocity or the position at a certain time, then you can use it to work out what the constant is.

Example:

The acceleration of an object is given by $\boldsymbol{a} = \begin{bmatrix} 4 \\ -2 \end{bmatrix}$ and its initial velocity is $\begin{bmatrix} 1 \\ 3 \end{bmatrix}$.

Find the velocity at time t.

$$\boldsymbol{v} = \int \boldsymbol{a}\, dt = \begin{bmatrix} 4t + c_1 \\ -2t + c_2 \end{bmatrix}$$

c_1 and c_2 are the constants of integration

When t = 0, $\boldsymbol{v} = \begin{bmatrix} 4\times 0 + c_1 \\ -2\times 0 + c_2 \end{bmatrix} = \begin{bmatrix} 1 \\ 3 \end{bmatrix}$

You were told what the initial velocity was in the question.

So $c_1 = 1$ and $c_2 = 3$.

You already worked out that $\boldsymbol{v} = \begin{bmatrix} 4t + c_1 \\ -2t + c_2 \end{bmatrix}$

Now just plug the constants in: $\boldsymbol{v} = \begin{bmatrix} 4t + 1 \\ -2t + 3 \end{bmatrix}$

Sample exam questions:

1) A particle moves in a straight line. At time t after starting its motion, its velocity is given by $v = t^2(15 - 2t)$.

a) Find the time t when the acceleration is zero. [3 marks]

b) Find how far the particle has moved when t = 4. [4 marks]

2) A particle moves with velocity at time t given by $\boldsymbol{v} = \begin{bmatrix} 6t - 2 \\ 5 \end{bmatrix}$.

a) Given that the initial position vector of the particle is $\begin{bmatrix} 4 \\ 3 \end{bmatrix}$, find the position vector for the particle at time t. [5 marks]

b) Show that the acceleration of the particle is constant. [2 marks]

I've got variable acceleration if I tap my feet to music when I'm driving...

Examiners are usually pretty predictable (no offence). Questions on this topic usually follow the same pattern: differentiate or integrate, work out any constants and then maybe work out a few values for particular times. Look out for the initial conditions (initial velocity or position) and remember the order of differentiation: position; velocity; acceleration.

Vectors

'Vector' might sound like a really dull Bond villain, but... well, it's not. Vectors have both size (or magnitude) and direction. If a measurement just has size but not direction, it's called a scalar.

- So, Vector, you expect me to talk?
- No, Mr Bond, I expect you to die! Ak ak ak!

Vectors have a Magnitude — Scalars Don't

Examples of vectors: velocity, displacement, acceleration, force

Examples of scalars: speed, distance

A really important thing to remember is that an object's speed and velocity aren't always the same:

Example: A runner sprints 100 m along a track at a speed of 8 ms^{-1} and then she jogs back 50 m at 4 ms^{-1}.

Average Speed

Speed = Distance ÷ Time

The runner takes (100 ÷ 8) + (50 ÷ 4) = 25 s to travel **150 m**.

So the average speed is 150 ÷ 25 = 6 ms^{-1}

Average Velocity

Velocity = Change in displacement ÷ Time

In total, the runner ends up **50 m** away from her start point and it takes 25 s.

So the average velocity is 50 ÷ 25 = 2 ms^{-1} in the direction of the sprint.

She jogged back 50 m after she jogged forward 100 m.

The Length of the Arrow shows the Magnitude of a Vector

You can add vectors together by drawing the arrows nose to tail.
The single vector that goes from the start to the end of the vectors is called the resultant vector.

It doesn't matter which order you draw the vectors.

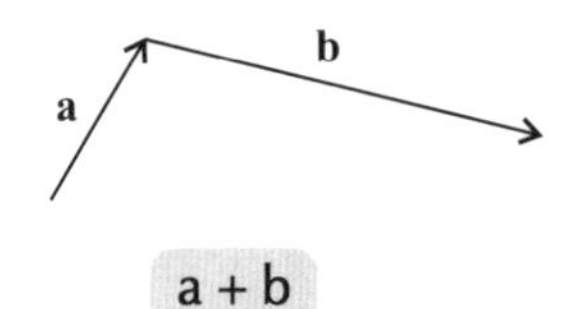

a + b

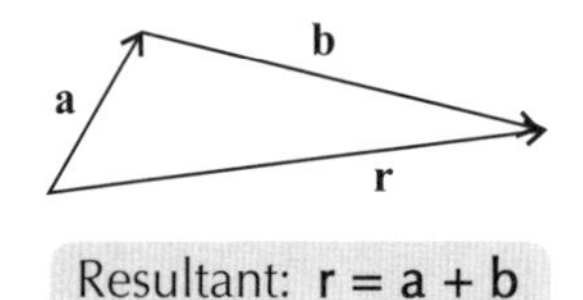

Resultant: **r = a + b**

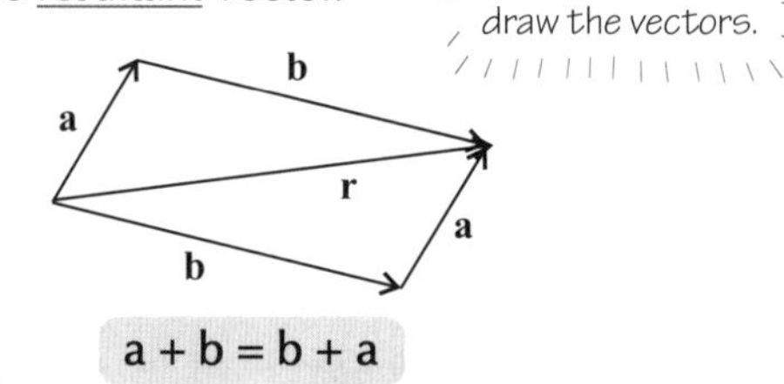

a + b = b + a

You can also multiply a vector by a scalar (just a number): the length changes but the direction stays the same.

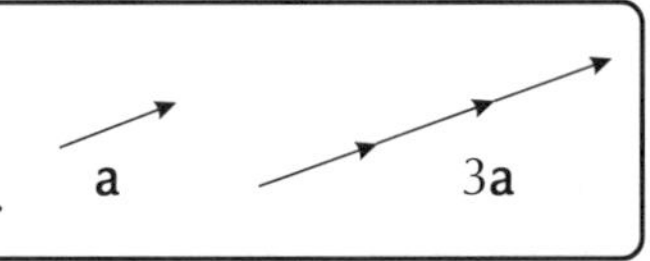

Resolving means writing a vector as Component Vectors

Splitting a vector up like this means you can work things out with one component at a time.
The unit vectors **i** and **j** are often used for resolving. They're called unit vectors because they each have a magnitude of 1. **i** is in the direction of the x-axis and **j** is in the direction of the y-axis.

Example: $\overrightarrow{AB} = 3\mathbf{i} + 2\mathbf{j}$ and $\overrightarrow{BC} = 5\mathbf{i} - 3\mathbf{j}$. Work out $\overrightarrow{AC}$.

Add the horizontal and vertical components separately.

$$\overrightarrow{AC} = \overrightarrow{AB} + \overrightarrow{BC} = (3\mathbf{i} + 2\mathbf{j}) + (5\mathbf{i} - 3\mathbf{j}) = 8\mathbf{i} - \mathbf{j}$$

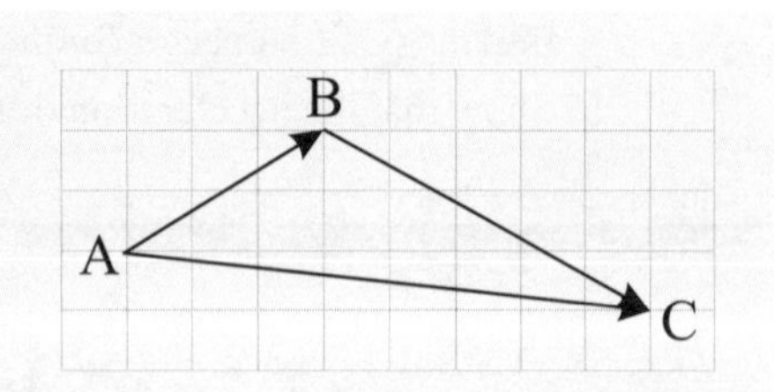

OCR B and AQA A only

Vectors can also be written as two components in a column vector:

$$\overrightarrow{AC} = \overrightarrow{AB} + \overrightarrow{BC} = \begin{bmatrix} 3 \\ 2 \end{bmatrix} + \begin{bmatrix} 5 \\ -3 \end{bmatrix} = \begin{bmatrix} 8 \\ -1 \end{bmatrix}$$

OCR B and AQA B only

A third unit vector, **k**, in the direction of the z-axis ('out of' the page), can be used for vectors in 3-D.

Vectors

Use Trig and Pythagoras to Change a vector into Component Form

Example: A ball travels with speed 5 ms^{-1} at an angle of 30° to the horizontal. Find the horizontal and vertical components of the ball's velocity, **v**.

First, draw a diagram and make a right-angle triangle:

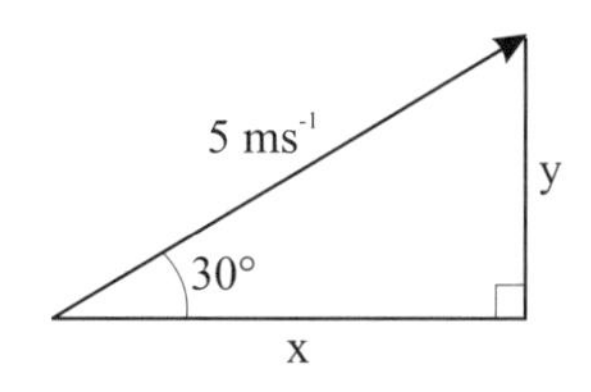

Using trigonometry, we can find x and y:

$$\cos 30° = \frac{x}{5} \quad \text{so } x = 5\cos 30°$$

$$\sin 30° = \frac{y}{5} \quad \text{so } y = 5\sin 30°$$

So $\mathbf{v} = (5\cos 30°\ \mathbf{i} + 5\sin 30°\ \mathbf{j})\ ms^{-1}$

Example: The acceleration of a body is given by the vector $\mathbf{a} = 6\mathbf{i} - 2\mathbf{j}$. Find the magnitude and direction of the acceleration.

Start with a diagram again. Remember, the y-component "-2" means "down 2".

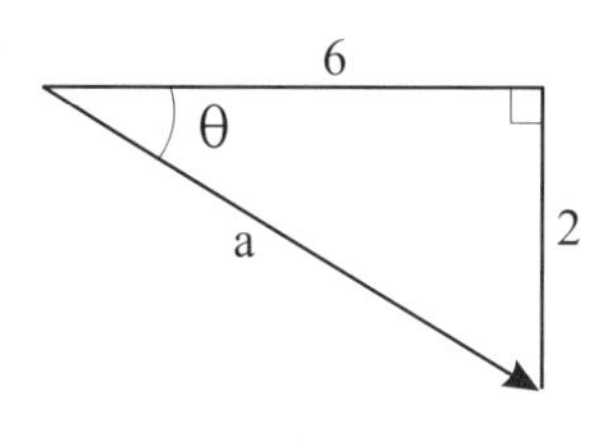

Using Pythagoras' Theorem, you can work out the magnitude of **a**:

$$a^2 = 6^2 + (-2)^2 = 40$$

$$\text{so } a = \sqrt{40} = 6.3 \text{ (2 sf)}$$

Use trigonometry to work out the angle:

$$\tan\theta = \frac{2}{6} \quad \text{so} \quad \theta = \tan^{-1}\frac{2}{6} = 18.4°$$

So vector **a** has magnitude 6.3 and direction 18.4° below the horizontal.

> In general, a vector with magnitude r and direction θ can be written as $r\cos\theta\mathbf{i} + r\sin\theta\mathbf{j}$
>
> The vector $x\mathbf{i} + y\mathbf{j}$ has magnitude $r = \sqrt{x^2 + y^2}$ and direction $\theta = \tan^{-1}\left(\frac{y}{x}\right)$

You can Resolve in any two Perpendicular Directions — not just x and y

Example: Find the resultant of the forces shown in the diagram.

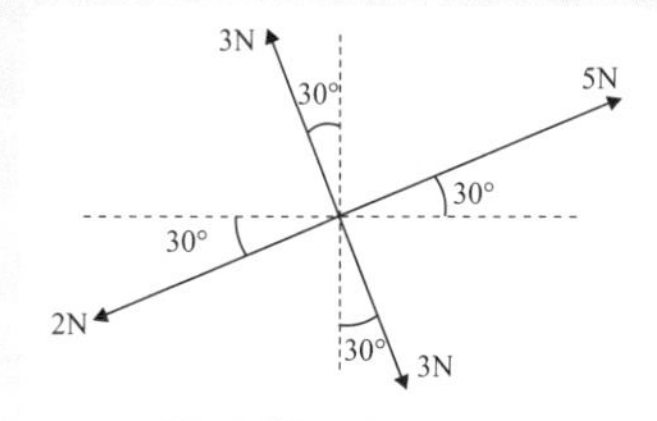

Resolving in ↖ direction: 3 N – 3 N = 0

Resolving in ↗ direction: 5 N – 2 N = 3 N

The two forces of 3 N balance each other.

So the resultant force is 3 N in the direction of the force of 5 N.

Practice Questions

1) ***Find the average velocity of a cyclist who cycles at 15 kmh^{-1} north for 15 minutes and then cycles south at 10 km/h for 45 minutes.***

2) ***Find a + 2b – 3c where a = 3i + 7j; b = -2i + 2j; c = i – 3j.***

3) ***Sample exam question:***

The diagram shows two forces acting on a particle. Find the magnitude and direction of the resultant force. [5 marks]

4N
30°
5N

Bet you can't say 'perpendicular' 10 times fast...

Next time you're baffled by vectors, just start resolving and Bob's your mother's brother.

Vector Motion

Skip these two pages if you're doing OCR A.

Vectors are much more than just a pretty face (or arrow). Once you deal with all the waffle in the question, all sorts of problems involving displacement, velocity, acceleration and forces can be solved using vectors.

Draw a Diagram if there's Lots of Vectors floating around

In fact, draw a diagram even when there's only a couple of vectors. But it's vital when there are lots of the little beggars.

Example: A ship travels 100 km at a bearing of 025°, then 75 km at 140° before going 125 km at 215°. What is the displacement of the ship from its starting point?

Remember that bearings are always measured starting from north.

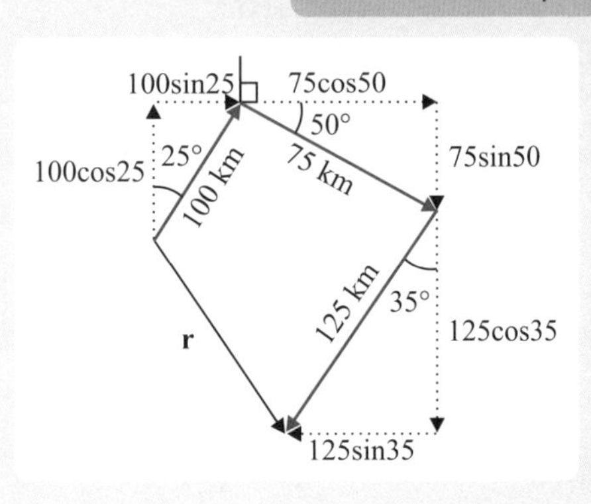

Resolve East: 100sin25 + 75cos50 – 125sin35 = 18.8 km

Resolve North: 100cos25 – 75sin50 – 125cos35 = -69.2 km

Magnitude of **r** = $\sqrt{18.8^2 + (-69.2)^2}$ = 71.7 km

direction $\theta = \tan^{-1}\left(\frac{69.2}{18.8}\right)$ = 74.8°

Bearing is 90° + 74.8° = 164.8°

So the displacement is 71.7 km on a bearing of 164.8°

The Direction part of a vector is Really Important

...and that means that you've got to make sure your diagram is spot on.

These two problems look similar, and the final answers are pretty similar too.
But look closely at the diagrams and you will see they are a bit different.

Example: A canoe is paddled at 4 ms^{-1} in a direction perpendicular to the seashore. The sea current has a velocity of 1 ms^{-1} parallel to the shore. Find the resultant velocity **r** of the canoe.

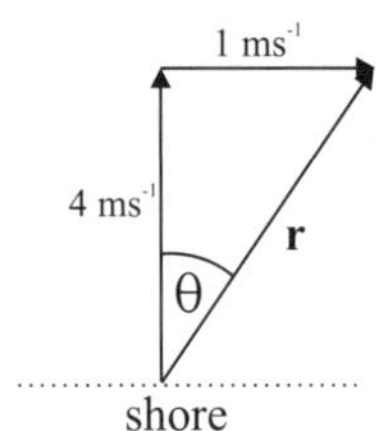

The resultant velocity **r** is the hypotenuse of the right-angle triangle.

Magnitude of **r** = $\sqrt{4^2 + 1^2}$ = 4.1 ms^{-1}

Direction: $\theta = \tan^{-1}\left(\frac{1}{4}\right)$ = 14.0°

Example: A canoe can be paddled at 4 ms^{-1} in still water. The sea current has a velocity of 1 ms^{-1} parallel to the shore. Find the angle at which the canoe must be paddled in order to travel in a direction perpendicular to the shore and the magnitude of the resultant velocity.

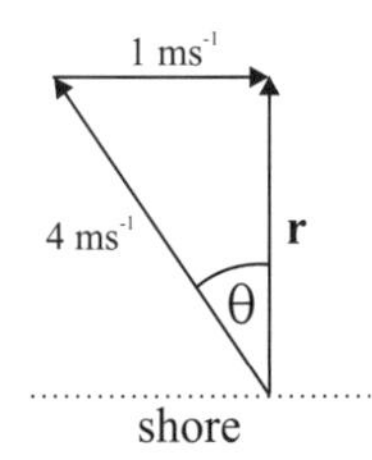

The resultant velocity **r** isn't the hypotenuse this time.

Magnitude of **r** = $\sqrt{4^2 - 1^2}$ = 3.9 ms^{-1} ⟸ The magnitude is different from the example above.

Direction: $\theta = \sin^{-1}\left(\frac{1}{4}\right)$ = 14.5°

Vector Motion

First Decide which vectors you need to Work Out

It's no good ploughing into a question if you're working out the wrong value. Once again, the diagram's the key.

Example:

A sledge of weight 1000 N is being held on a rough slope at an angle of 35° by a force parallel to the slope of 700 N. Find the normal contact force N and the frictional force F acting on the sledge.

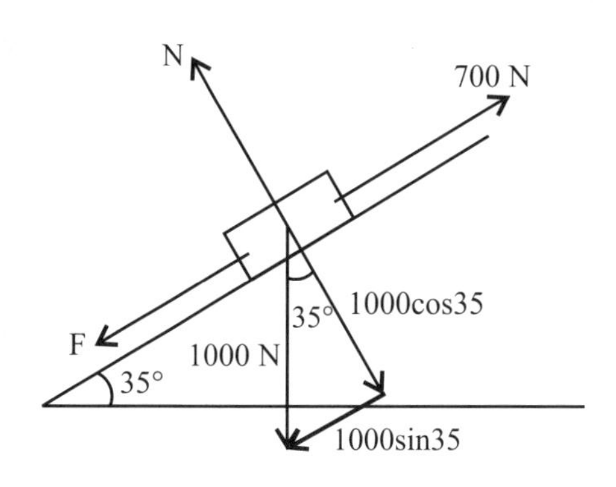

All the forces involved here act either parallel or perpendicular to the slope so it makes sense to resolve in these directions.

Perpendicular to the slope:

$N - 1000\cos 35 = 0$
So $N = 1000\cos 35 \quad = 819.2 \text{ N}$

Parallel to the slope:

$700 = 1000\sin 35 + F$
So $F = 700 - 1000\sin 35° \quad = 126.4 \text{ N}$

Practice Questions

1) ***A plane flies 40 miles due south, then 60 miles southeast before going 70 miles on a bearing of 020°. Find the distance and bearing on which the plane must fly to return to its starting point.***

2) ***A toy train of weight 25 N is pulled up a slope of 20°. The tension in the string is 25 N and a frictional force of 5 N acts on the train. Find the normal contact force and the resultant force acting on the train.***

3) ***Sample exam question:***

A girl can swim at 2 ms^{-1} in still water. She is swimming across a river in a direction perpendicular to the riverbank. The river is flowing at 3 ms^{-1} and so it carries the girl downstream.
Find the magnitude of the resultant velocity of the girl and the angle it makes with the riverbank. [6 marks]

'Vector motion' is an anagram of 'croon, vote tim'...

Diagrams are really important and really useful for solving problems using vectors. Sketch out a nice clear diagram before you decide how to tackle the problem. There's usually loads of different things you could work out, so always have a check back at the end to make sure you have worked out what the question was actually asking you for.

Mathematical Modelling

You'll have to make lots of assumptions in M1. Doing this is called 'modelling', and you do it to make a sticky real-life situation simpler.

Hint: 'modelling' in maths doesn't have anything to do with plastic aeroplane kits... or catwalks.

Example: **The ice hockey player**

You might have to assume:

- no friction between the skates and the ice
- no drag (air resistance)
- the skater generates a constant forward force S
- the skater is very small (a point mass)
- there is only one point of contact with the ice
- the weight acts downwards

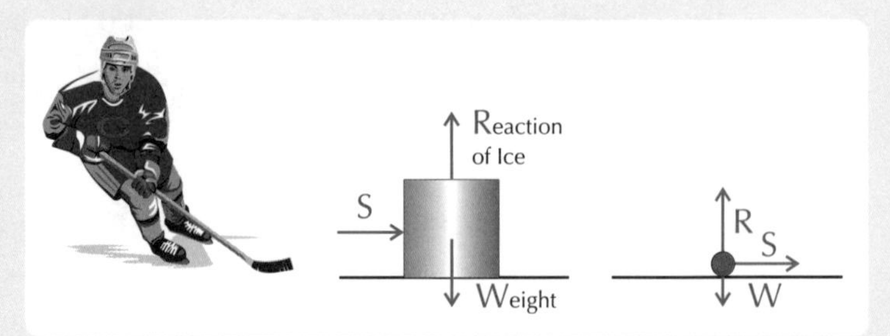

Wayne Grotski — ice hockey superstar

Wayne as a block, with forces shown

Wayne as a point mass, with forces

The complex hockey player on the left has become a simple mathematical model on the right with only three forces. Easy.

Modelling is a Cycle

Having created a model you can later improve it by making more (or less) assumptions.

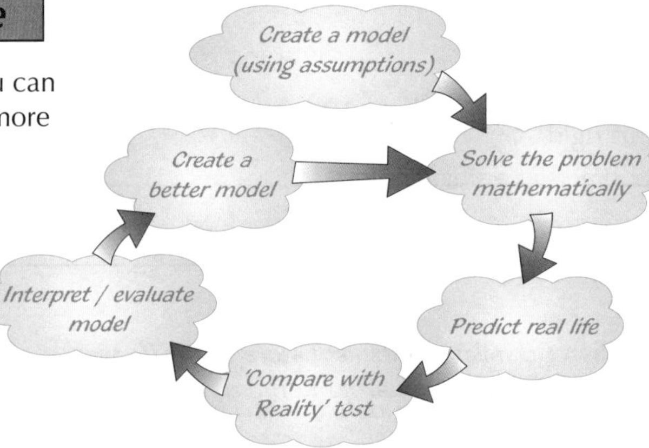

Mmm... green and red. I think this graphic wins the coveted Award for Most Tasteless Colour Scheme.

Talk the Talk

Maths questions in M1 use a lot of words that you already know, but here they're used to mean something very precise. Learn these definitions so you don't get caught out:

Particle — the 'mass' or 'body' drawn as a rectangle or dot (e.g. Wayne the hockey player)
Light — the body has no mass (e.g. a feather)
Smooth — the surface doesn't have friction / drag opposing motion (e.g. an ice rink)
Rough — the surface will oppose motion with friction / drag (e.g. a table top)
Beam or Rod — a long particle (e.g. a carpenter's plank)
Uniform — the mass is evenly spread out throughout the body (e.g. a school ruler)
Non-uniform — the mass is unevenly spread out (e.g. along a tennis racket)
Rigid — the body does not bend (e.g. a metal ruler)
Thin — the body has no thickness
Lamina — a surface that is thin (e.g. a sheet of A4 paper)
Equilibrium — nothing's moving
Plane — a flat surface (e.g. a table top)
Tension — the force in a taut wire, rope or string
Inextensible — the body can't be stretched (e.g. a metal rod)
Static — not moving

Mathematical Modelling

Always start by drawing a ***Simple Diagram*** *of the* ***Model***

Here's a couple of old chestnuts that often turn up in M1 exams in one form or another.

Example: **The book on a table**

A book is put flat on a table. One end of the table is slowly lifted and the angle to the horizontal is measured when the book starts to slide. What assumptions might you make?

Assumptions:
The book is rigid, so it doesn't bend or open.
The book is a particle, so its dimensions don't matter.
There's no wind or other external forces involved.

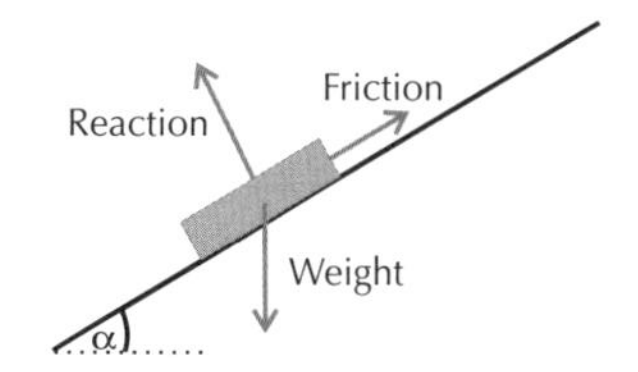

Example: **The balance**

A pencil is placed on a table and a 12″ ruler is put across it so that it balances. A 1p coin is placed on one side and a 10p coin on the other so that the ruler still balances. Draw a model of the forces. What assumptions have you made?

Assumptions:
The coins are point masses.
The ruler is rigid.
The support acts at a single point.

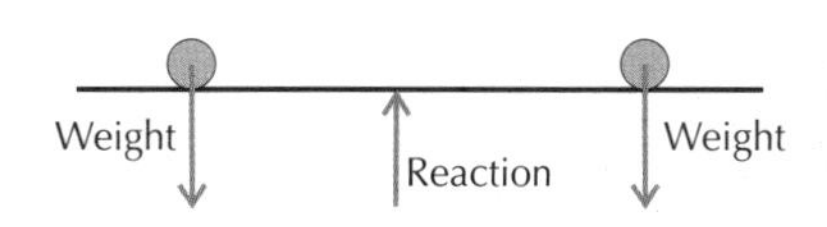

Practice Questions

1) The following items are dropped from a height of 2 m onto a cushion:

a) a full 330ml drinks can b) an empty drinks can c) a table tennis ball

The time each takes to fall is measured.

Draw a model of each situation and list any assumptions which you've made.

2) A car is travelling at 25 mph along a level road.

Draw a model of the situation and list any assumptions which you've made.

3) A skydiver is falling to earth before his parachute opens.

Draw a model of the situation and list any assumptions which you've made.

Phobia #1 — pteronophobia: fear of being tickled with feathers...

I've said it before and I'll say it again: if it makes their lives easier, examiners always stick the same kinds of questions into M1 Exams year after year. If you practise enough 'books on tables' and 'balancing pencil' questions, you'll have no nasty surprises in the Exam. And if you feel you've just learned more about ice hockey then you've missed the point slightly.

Forces are Vectors

Forces have magnitude and direction. Only force arrows should be attached to a particle. Geddit? Gottit. Good.

Forces have *Components*

You've done a fair amount of trigonometry already, so this should be as straightforward as watching dry paint.

Example: A particle is acted on by a force of 15 N at 30° above the horizontal. Find the horizontal and vertical components of the force.

A bit of trigonometry is all that's required:

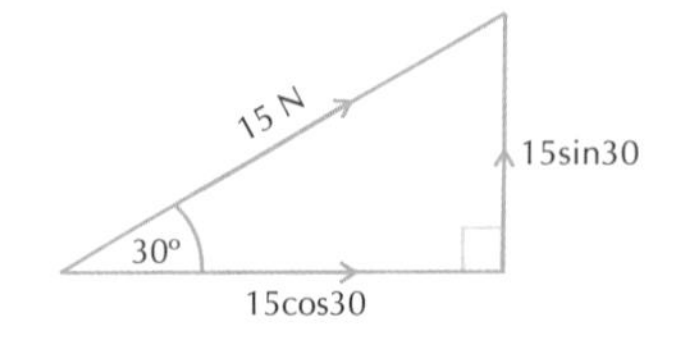

$$\text{Force} = \begin{pmatrix} 15\cos 30° \\ 15\sin 30° \end{pmatrix}$$
$$= (13.0\,\mathbf{i} + 7.5\,\mathbf{j})\ \text{N}$$

(i.e. 13 N to the right and 7.5 N upwards)

Add Forces Top to Tail to get the *Resultant*

The important bit when you're drawing a diagram to find the resultant is to make sure the arrows are the right way round. Repeat after me: top to tail, top to tail, top to tail.

Example: A second horizontal force of 20 N is also applied to the particle in the example above. Find the resultant of these forces.

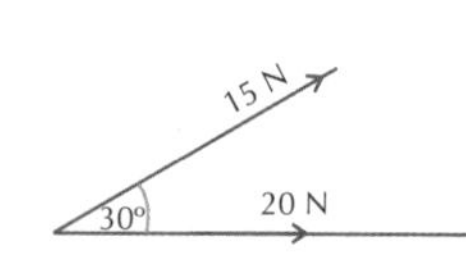

You need to put the arrows top to tail:

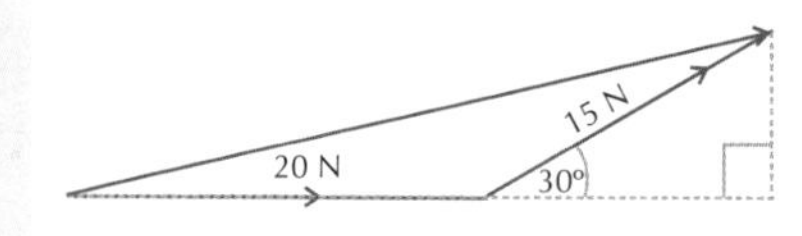

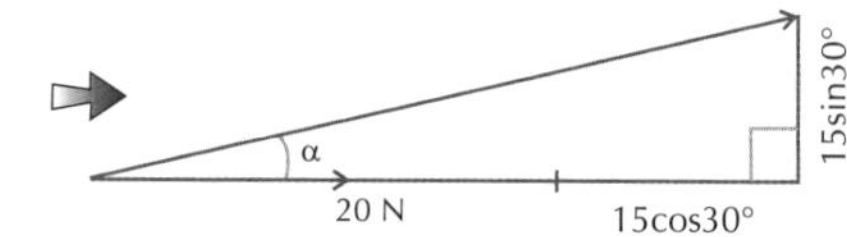

Using Pythagoras and trigonometry:

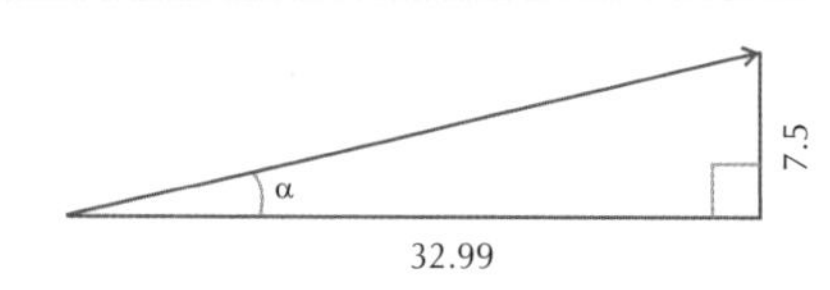

$$R = \sqrt{32.99^2 + 7.5^2} = 33.8\ \text{N}$$

$$\alpha = \tan^{-1}\left(\frac{7.5}{32.99}\right) = 12.8°\ \text{above the horizontal}$$

Example: Find the magnitude and direction of the resultant of the forces shown acting on the particle.

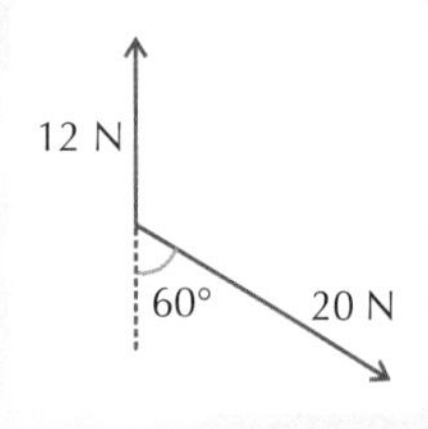

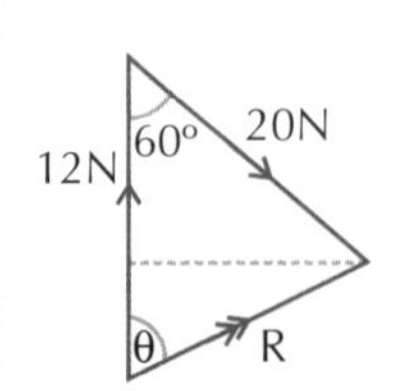

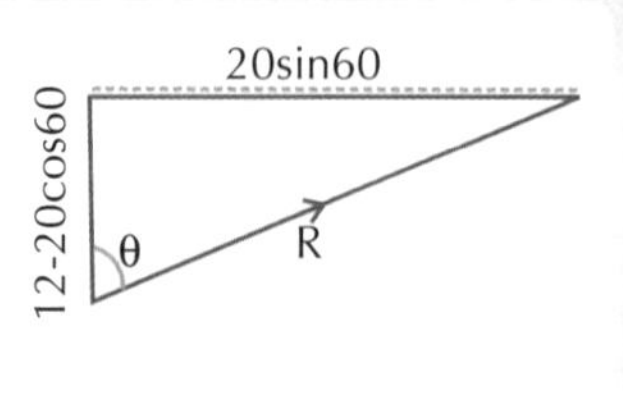

Hint: you could use the cosine rule here to get R.

$$R = \sqrt{(12 - 20\cos 60°)^2 + (20\sin 60°)^2} = 17.4\ \text{N}$$

$$\theta = \tan^{-1}\left(\frac{20\sin 60°}{12 - 20\cos 60°}\right) = 83.4°\ \text{to vertical}$$

If a particle is released it will move in the direction of the resultant.

Forces are Vectors

Particles in **Equilibrium Don't Move**

Forces acting on a particle in equilibrium add up to zero force. That means when you draw all the arrows top to tail, you finish up where you started. That's why diagrams showing equilibrium are called 'polygons of forces'.

Example: Find the magnitude of force P for equilibrium.

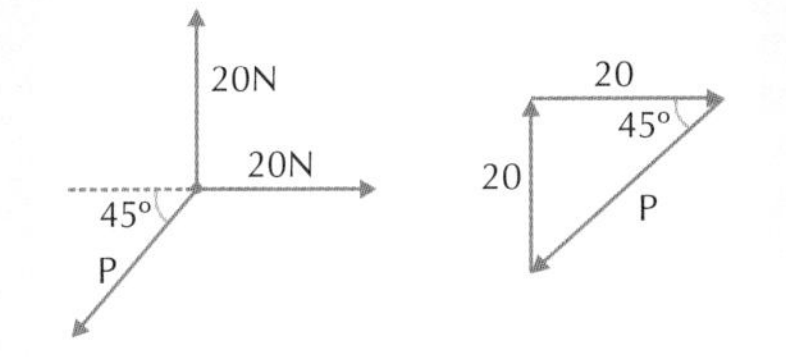

$$P\cos 45 = 20$$
$$P = 28.3 \text{ N}$$

Example: Find S and T for equilibrium.

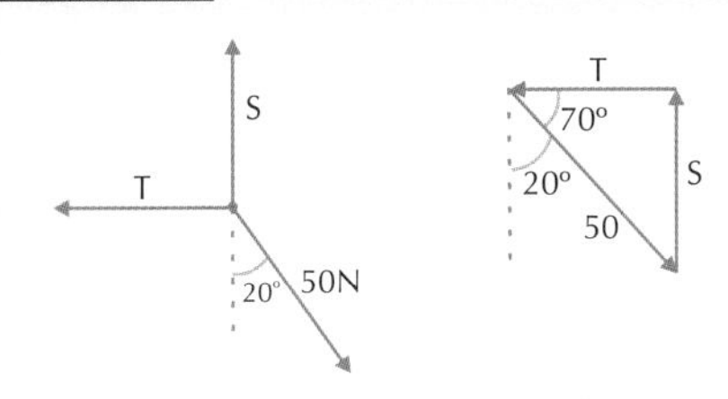

$$S = 50\sin 70° = 47.0 \text{ N}$$
$$T = 50\cos 70° = 17.1 \text{ N}$$

Example: Find P and Q for equilibrium.

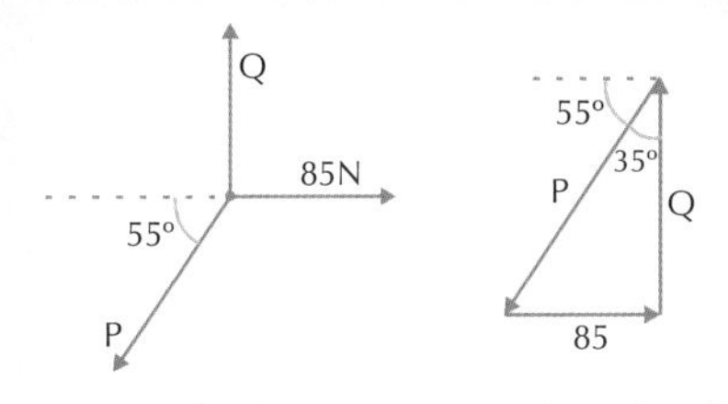

$$\sin 35° = \frac{85}{P} \quad \text{so } P = 148 \text{ N}$$
$$\tan 35° = \frac{85}{Q} \quad \text{so } Q = 121 \text{ N}$$

Practice Questions

1) Find the magnitudes and directions to the horizontal of the resultant force in each situation.

a)

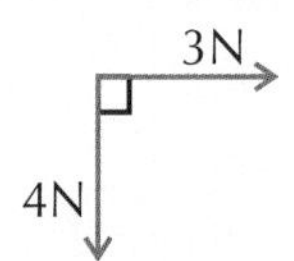

b)

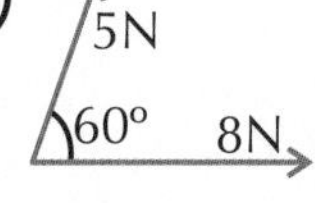

c)

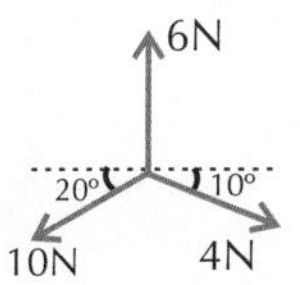

Sample exam questions:

2) A force of magnitude 7 N acts horizontally on a particle. Another force, of magnitude 4 N, acts on the particle at an angle of 30° to the horizontal. The resultant of the two forces has a magnitude R at an angle α to the horizontal.

Find a) The force R [3 marks]

b) The angle α [3 marks]

3) Three forces of magnitudes 15 N, 12 N and W act on a particle as shown.
Given that the particle is in equilibrium, find:

a) The value of θ [2 marks]

b) The force W [2 marks]

The force W is now removed.
State the magnitude and direction of the resultant of the two remaining forces. [2 marks]

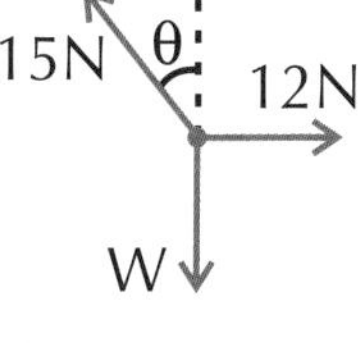

Phobia #2 — coulrophobia: fear of clowns...

...the red noses, the baggy pants, the pratfalls... uh. Better have a lie down and forget all about it.

Types of Forces

Different types of forces act on a body for different reasons...

Weight (W)

Due to the particle's mass, m and the force of gravity, g: **W = mg** — weight always acts downwards.

The Normal Reaction (R or N)

The reaction from a surface. Reaction is always at 90° to the surface.

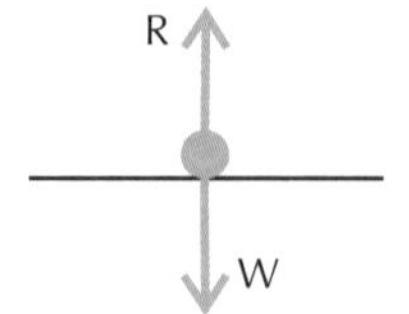

Tension (T)

Force in a taut rope, wire or string.

Friction (F)

Due to the roughness between a body and a surface. Always acts against motion, or likely motion.

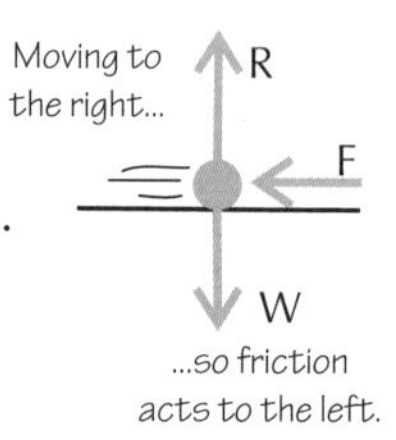

Thrust

(If you're doing OCR A or AQA A you don't need this bit.)

Force in a rod (e.g. the pole of an open umbrella).

Example: A sledge is being steadily pulled by a small child on horizontal snow. Draw a force diagram for a model of the sledge. List your assumptions.

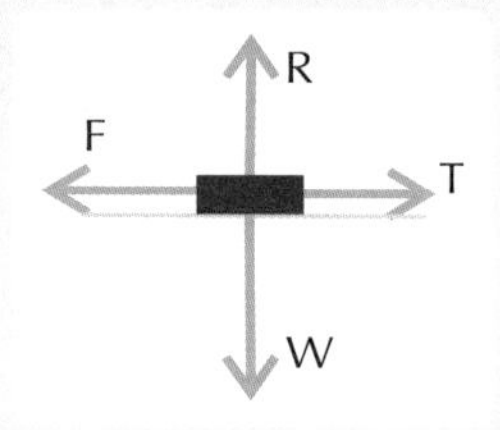

Assumptions:

1) Friction is too big to be ignored (i.e. it's not ice).
2) The string is horizontal (it's a small child).
3) Take the sledge to be a small particle (so its size doesn't matter).

Example: A mass of 12 kg is held by two light strings, P and Q, acting at 40° and 20° to the vertical as shown. Find the tension in each string. Take g = 9.8 ms^{-2}.

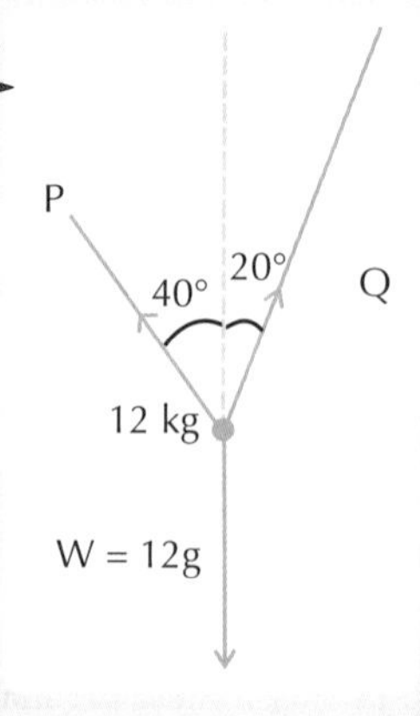

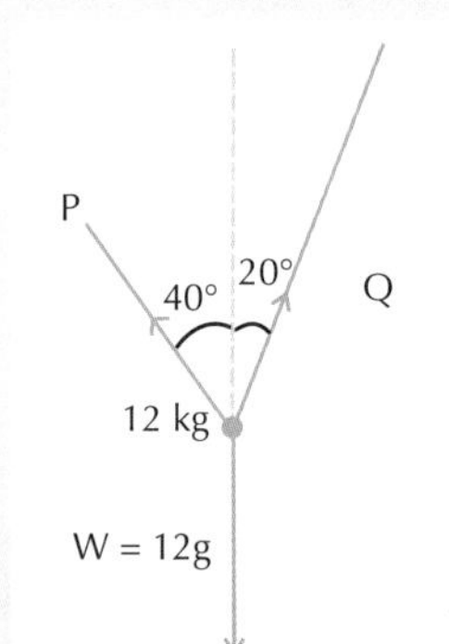

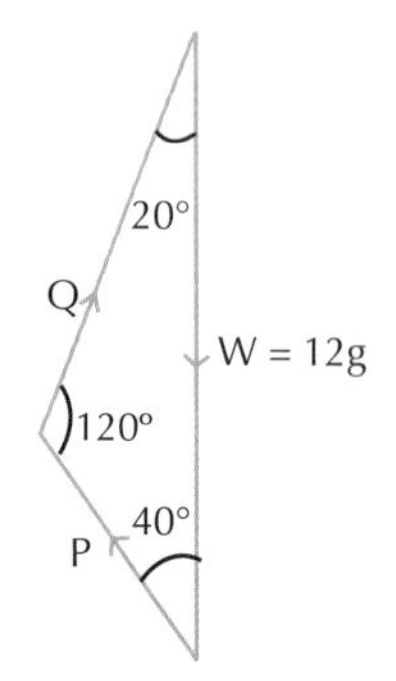

Sine rule:

$$\frac{P}{\sin 20} = \frac{12g}{\sin 120}$$

So P = 46.4 N

$$\frac{Q}{\sin 40} = \frac{12g}{\sin 120}$$

So Q = 87.3 N

Always look out for sine rule triangles in your polygons of forces.

Types of Forces

An Inclined Plane is a Sloping Surface

Example: A particle of mass 0.1 kg is held at rest on a rough plane inclined at 20° to the horizontal by a friction force acting up the plane. Find the magnitude of this friction force and the normal reaction. (Take g = 9.8 ms^{-2}.)

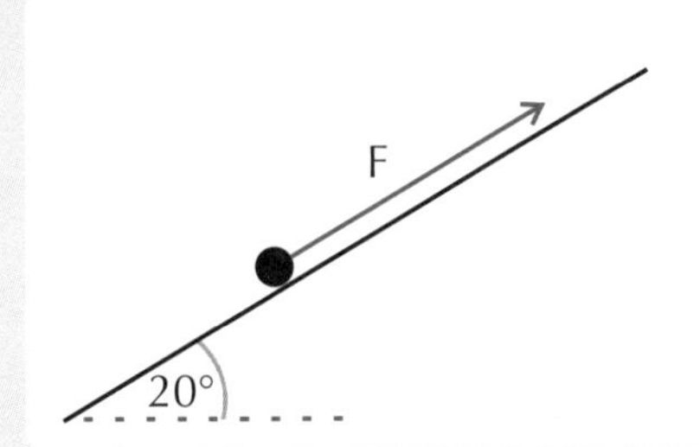

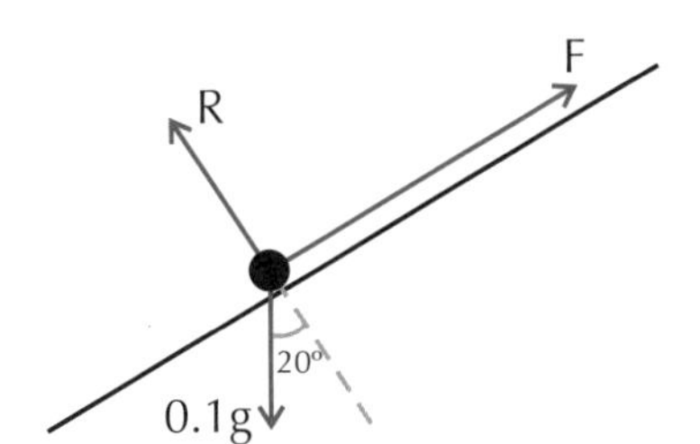

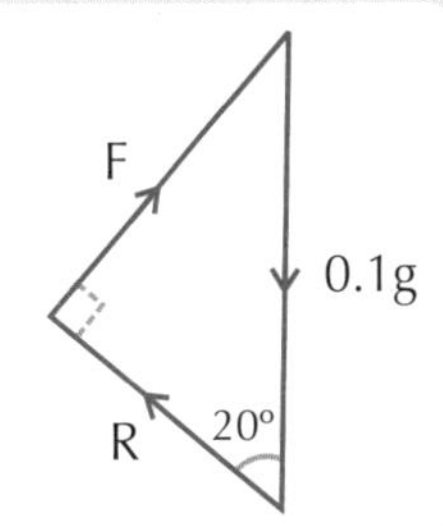

$F = 0.1g\sin 20°$
$= 0.335$ N
$R = 0.1g\cos 20°$
$= 0.921$ N

Practice Questions

Take g = 9.8 ms^{-2} in each of these questions.

1) ***A mass of M kg is suspended by two light wires A and B, with angles 60° and 30° to the vertical respectively, as shown. The tension in A is 20 N. Find:***
   a) the tension in wire B
   b) the mass M

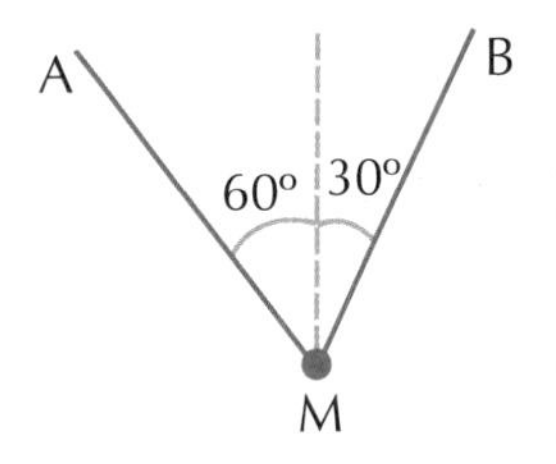

2) ***A particle, Q, of mass m kg, is in equilibrium on a smooth plane which makes an angle of 60° to the vertical. This is achieved by an attached string S, with tension 70 N, angled at 10° to the plane as shown. Draw a force diagram and find both the mass of Q and the reaction on it from the surface.***

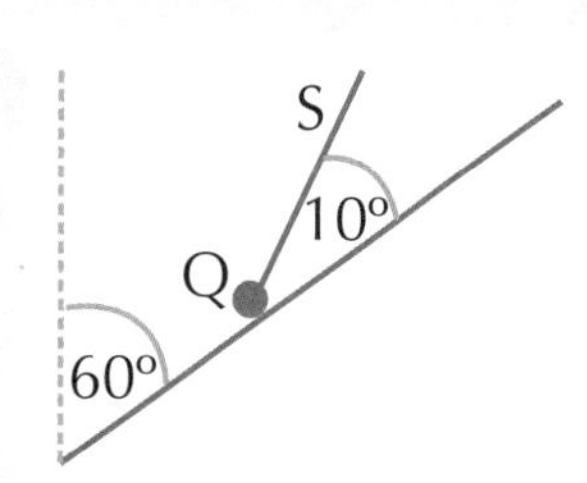

3) ***Sample exam question:***

A sledge is held at rest on a smooth slope which makes an angle of 25° with the horizontal. The rope is at an angle of 20° to the slope. It is given that the normal reaction acting on the sledge due to contact with the surface is 80 N.

Find:

a) The tension, T, in the rope. [3 marks]

b) The weight of the sledge. [3 marks]

Phobia #3 — pogonophobia: fear of beards...

Those 5 types of forces on the opposite page crop up loads in M1 Exams (except thrust, which isn't so common). You need to be completely familiar with all the jargon that gets bandied around in this subject. And it's absolutely vital that you can do inclined planes questions really well, because there's no way you'll get out of the Exam hall without getting one of those...

Friction

OCR B students don't need to do calculations involving the coefficient of friction until M2.

Friction Tries to **Prevent Motion**

Push hard enough and a particle will move, even though there's friction opposing the motion — so a friction force, F, has a maximum value. This depends on the roughness of the surface and the value of the normal reaction from the surface.

μ has no units. μ is pronounced as 'mu'.

μ is called the "coefficient of friction". The rougher the surface, the bigger μ gets.

Example:

What range of values can a friction force take in resisting a horizontal force P acting on a particle Q, of mass 12 kg, resting on a rough horizontal plane which has a coefficient of friction of 0.4? (Take $g = 9.9\ ms^{-2}$.)

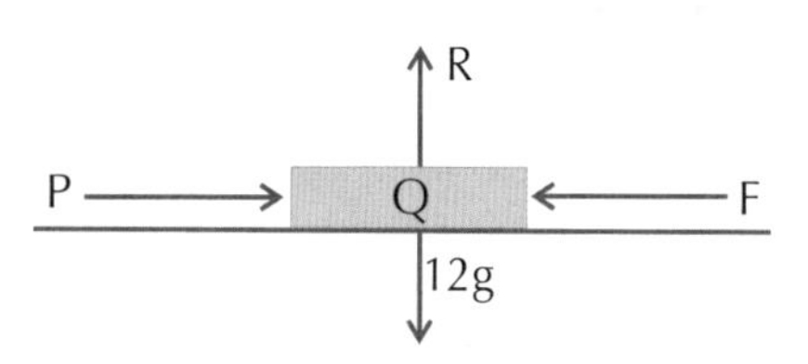

Resolving vertically: $R = 12g$

Use formula from above: $F \leq \mu R$

$F \leq 0.4(12g)$

$F \leq 47.04$ N

So friction can take any value between 0 and 47.04 N, depending on how large P is.

If $P \leq 47.04$ N then Q remains in equilibrium. If $P = 47.04$ N then Q is on the point of sliding — i.e. friction is at its limit. If $P > 47.04$ then Q will start to move.

Limiting Friction is When Friction is **Maximum** ($F = \mu R$)

Example:

A particle of mass 6 kg is placed on a rough horizontal plane which has a coefficient of friction 0.3. A horizontal force Q is applied to the mass. Describe what happens if Q is:

a) 16 N

b) 20 N

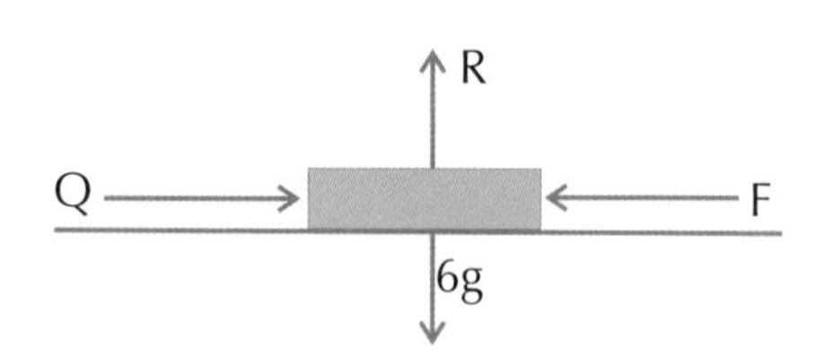

Resolving vertically: $R = 6g$

Using formula above: $F \leq \mu R$

$F \leq 0.3(6g)$

$F \leq 17.64$ N

a) Since $Q < 17.64$ it won't move.

b) Since $Q > 17.64$ it'll start moving. No probs.

Example:

A particle of mass 4 kg at rest on a rough horizontal plane is being pushed by a horizontal force of 30 N. Given that the particle is on the point of moving, find the coefficient of friction.

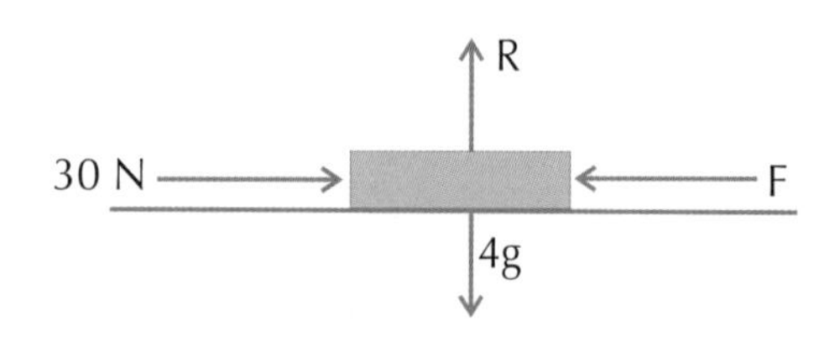

Resolving horizontally: $F = 30$

Resolving vertically: $R = 4g$

The particle's about to move, so friction is at its limit:

$F = \mu R$

$30 = \mu(4g)$

$$\mu = \frac{30}{4g} = 0.77$$

Friction

Example: Look back to the example on p 17. Given that the mass is only just held in equilibrium (i.e. it's about to slide down the plane), find μ.

Since it's limiting friction: $F = \mu R$

Therefore: $0.1g\sin 20° = \mu(0.1g\cos 20°)$

So: $\mu = \frac{0.1g\sin 20°}{0.1g\cos 20°} = \tan 20° = 0.36$

Both the 0.1 mass and g cancel — to find μ you didn't need to know the mass.

In fact, when a particle is about to slide on a rough plane inclined at α to the horizontal, and there are no other forces involved other than W (the weight), F and R, then **μ = tan α**.

Practice Questions

1) ***Describe the motion of a mass of 12 kg pushed by a force of 50 N parallel to the rough horizontal plane on which the mass is placed. The plane has coefficient of friction*** $\mu = \frac{1}{2}$***. (Take g = 9.8 ms***$^{-2}$***.)***

2) ***What minimum force would be needed to move the mass in Q1?***

Sample exam questions:

3) A particle is placed on a rough inclined plane with coefficient of friction μ = 0.2. At what angle to the horizontal is the plane, if the particle is about to slide? [2 marks]

4)

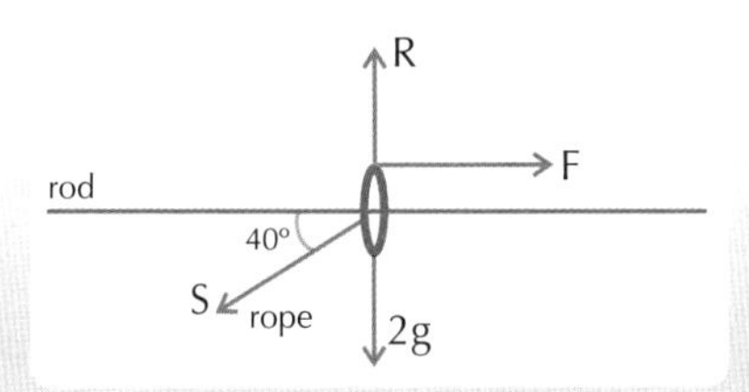

A 2 kg ring threaded on a rough horizontal rod is pulled sideways by a rope having tension S, as shown. The coefficient of friction between the rod and the ring is $\frac{3}{10}$. Given that the ring is about to slide, find the magnitude of S. [5 marks]

Phobia #4 — xenoglossophobia: fear of foreign languages...

Friction is a pain in the posterior for two reasons:
1) it stops you being able to slide all the way home with your eyes shut after a long night out;
2) it makes M1 questions just that little bit more complicated.

Moments

Skip these two pages if you're doing OCR A, OCR B or AQA A.

Moments are **Clockwise** or **Anti-Clockwise**

A 'moment' is the turning effect a force has around a point.
The larger the force, and the greater the distance from a point, then the larger the moment.

Moment = Force × Perpendicular Distance

The units are just newtons × metres = Nm. Couldn't they have thought of a cleverer name?

In these examples you need to take moments about point O each time:

Example:

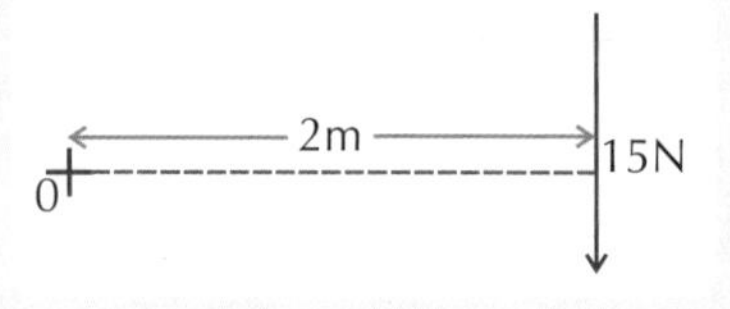

Moment = F × d
= 15 × 2
= 30 Nm

Example:

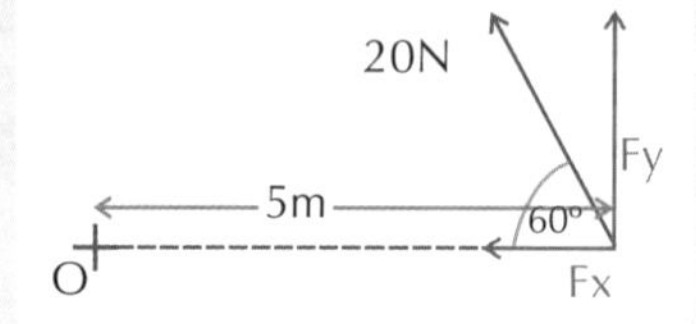

The 20 N force has components Fx and Fy.
Fx goes through O so its moment is zero.
Resolve vertically: Fy = 20sin60°
Moment = 20sin60° × 5
= 86.6 Nm

Example:

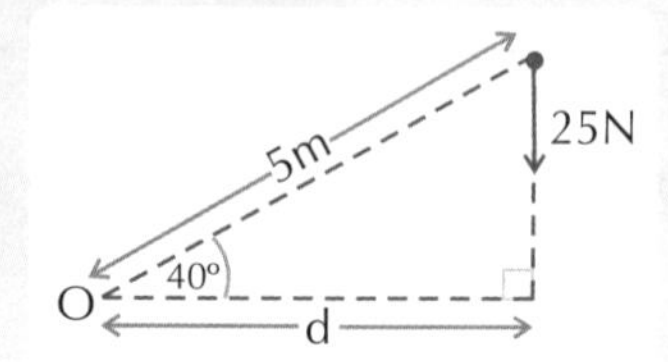

Remember it's got to be perpendicular distance, so you can't just plug in the 5m.
Here d is 5cos40°, so:
Moment = 25 × 5cos40°
= 95.8 Nm

In **Equilibrium** Moments Total **Zero**

...and that means that for a body in equilibrium the total moments either way must be equal:

Total Clockwise Moment = Total Anticlockwise Moment

Example: Two weights of 30 N and 45 N are placed on a light 8 m beam. The 30 N weight is at one end of the beam as shown whilst the other weight is a distance d from the midpoint M. The beam is in equilibrium held by a single wire with tension T attached at M. Find T and the distance d.

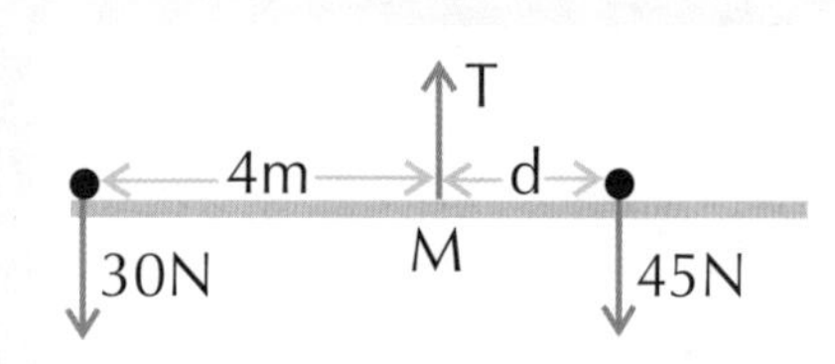

Resolving vertically: 30 + 45 = T = 75N
Take moments about M:
Clockwise Moment = Anticlockwise Moment
45 × d = 30 × 4

$$d = \frac{120}{45} = 2\frac{2}{3}\text{ m}$$

The **Weight** of a **Uniform Beam** Acts at its **Middle**

Example: A 6 m long uniform beam AB of weight 40 N is supported at A by a vertical reaction R. AB is held horizontal by a vertical wire attached 1 m from the other end. A weight of 30 N is placed 2 m from the support R. Find the tension T in the wire and the force R.

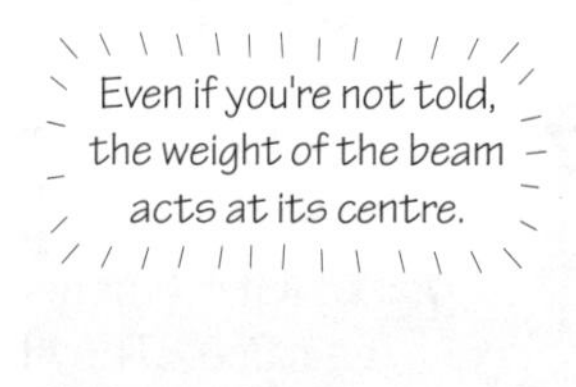

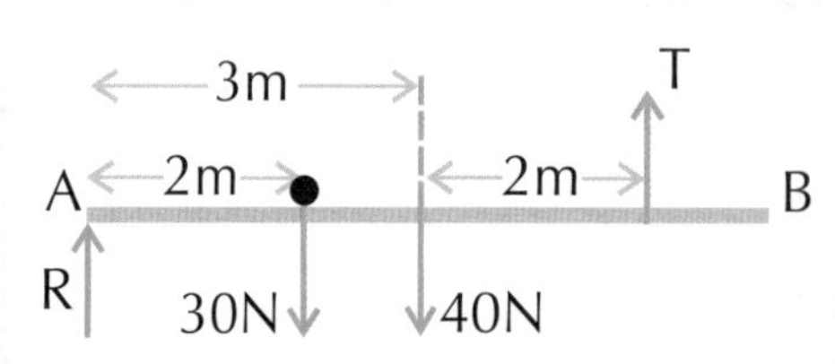

Take moments about A.
Clockwise Moment = Anticlockwise Moment
(30 × 2) + (40 × 3) = T × 5
T = 36 N
Resolve vertically: T + R = 30 + 40
So: R = 34 N

Moments

Take Moments Wisely

By taking moments about A in the last example you ended up with an equation containing only T.
That's because R goes through A, so has no moment about it.

It's always easier to do questions if you take moments about a point that has an unknown force going through it.

Example: ***If you're doing Edexcel you won't have to do 'leaning ladder' questions until M2.***

A uniform 12 kg ladder AB of length 6 m leans on a smooth vertical wall making an angle of 60° to the rough horizontal ground. Find the normal reactions R_1 and R_2 from the ground and the wall, given that a friction force F just prevents the ladder from slipping. What is the coefficient of friction between the ladder and the ground?

(Take g = ms^{-2})

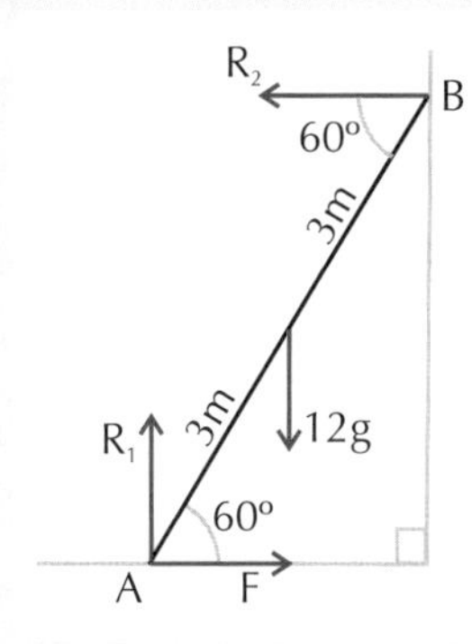

Resolving vertically: $R_1 = 12g = 117.6$ N

Moments about A: $12g \times 3\cos60° = R_2 \times 6\sin60°$

$R_2 = 33.9$ N

Resolving horizontally: $F = R_2$

$F = 33.9$ N

The ladder's on the point of slipping, so it's limiting friction:

$F = \mu R_1$

$33.9 = \mu \times 117.6$

$\mu = 0.288$

Practice Questions

1) ***A 60 kg uniform beam AE of length 14 m is in equilibrium, supported by two vertical ropes attached to B and D as shown.***

Find the tensions in the ropes to 1 d.p. Take g = ms^{-2}.

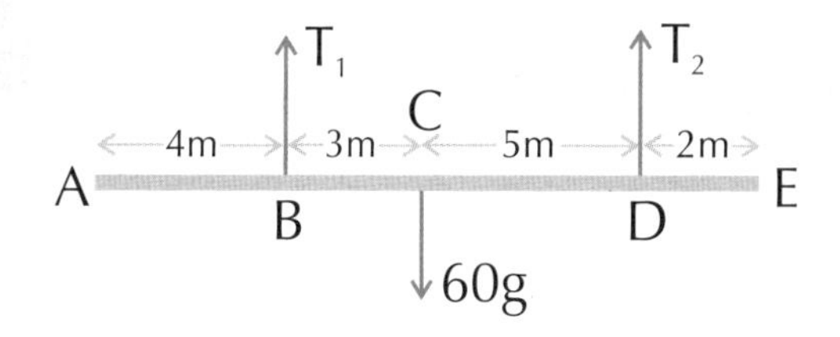

2) ***Sample exam question (skip if you're doing Edexcel):***

A uniform ladder of length 4 m has mass 15 kg. It rests in equilibrium against a smooth vertical wall making an angle of 75° to the rough horizontal ground. A 5 kg bucket of paint is attached three quarters of the way up the ladder.

Draw a diagram of the ladder including all forces acting on it. [2 marks]

Find:

a) the normal reaction from the wall; [3 marks]

b) the normal reaction from the ground; [2 marks]

c) the smallest value (to 2 d.p.) of the coefficient of friction between ground and ladder, if friction is limiting. [3 marks]

Phobia #5 — alliumphobia: fear of garlic...

Learn those formulas in the checked boxes — without knowing them you won't get very far with questions about moments. Whenever I go to an Italian restaurant I always order a Quattro Stagioni pizza, and M1 examiners are just the same. They'll almost always stick with what they know and include a question about ladders leaning against walls.

Centres of Mass

Skip these two pages if you're doing Edexcel, OCR A, OCR B or AQA A.

Working out a centre of mass, G, is all about working out how a structure will hang when it's suspended.

You can find out *Where G is* using *Symmetry*

Shapes with mass evenly distributed have G on their line(s) of symmetry.

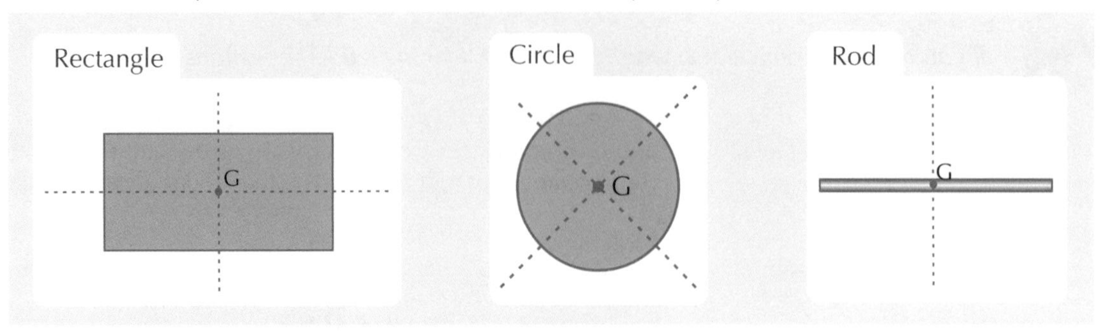

G Hangs Beneath Points of Suspension

Example: A rectangular lamina ABCD with dimensions as shown is suspended from B. What angle does AB make to the vertical?

A lamina is just a 2-D shape.

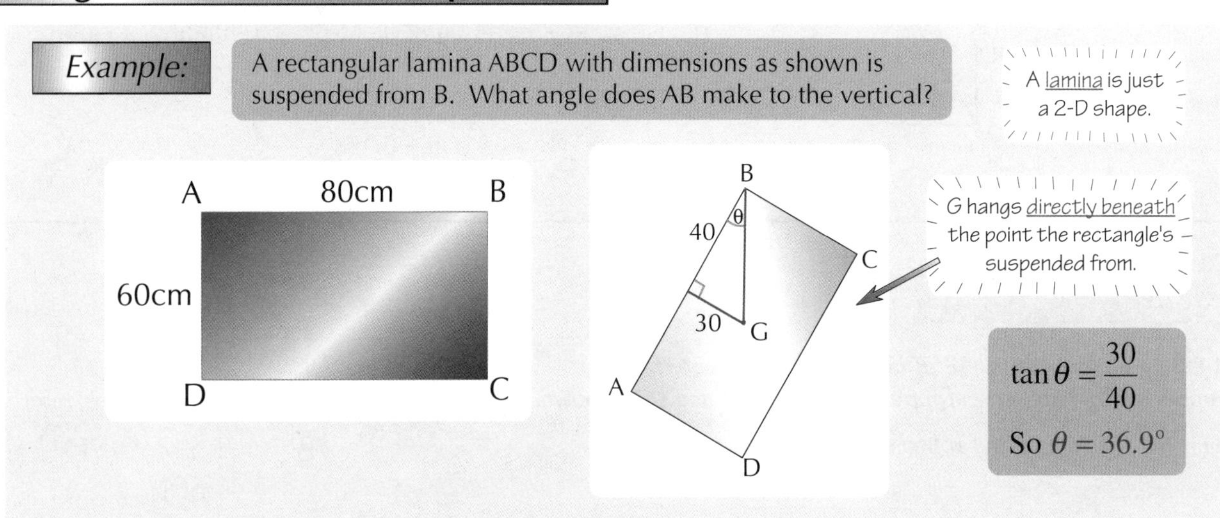

$\tan\theta = \frac{30}{40}$

So $\theta = 36.9°$

Combined Particles have a Total Mass, M

When two or more things are combined you need to use this formula:

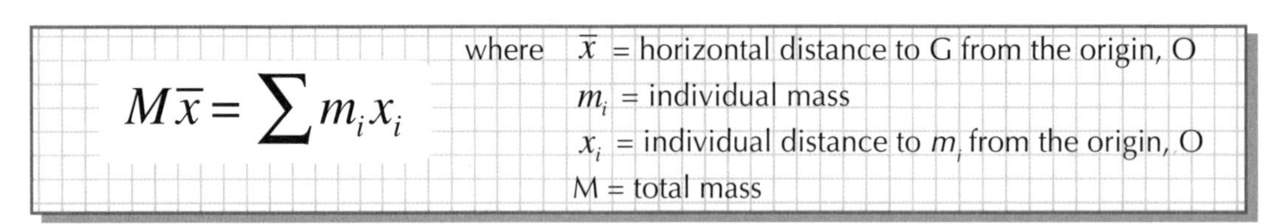

$$M\bar{x} = \sum m_i x_i$$

where $\bar{x}$ = horizontal distance to G from the origin, O

m_i = individual mass

x_i = individual distance to m_i from the origin, O

M = total mass

Example: Two uniform rods of lengths 3 m and 4 m with masses 6 kg and 12 kg respectively are joined in line as shown. Find the position of their combined centre of mass.

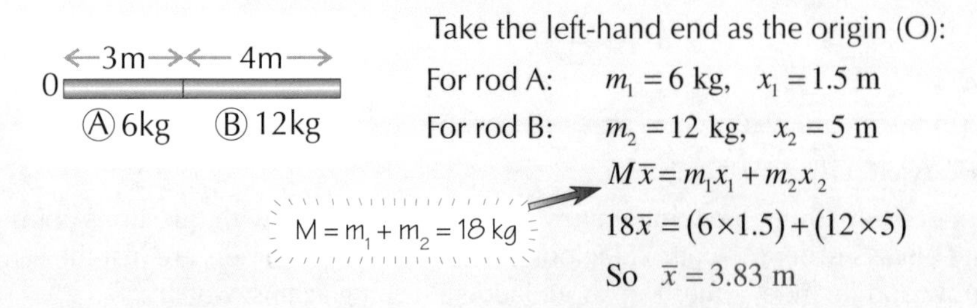

Take the left-hand end as the origin (O):

For rod A: $m_1 = 6$ kg, $x_1 = 1.5$ m

For rod B: $m_2 = 12$ kg, $x_2 = 5$ m

$M\bar{x} = m_1x_1 + m_2x_2$

$18\bar{x} = (6\times1.5) + (12\times5)$

So $\bar{x} = 3.83$ m

Centres of Mass

2D Structures need $M\bar{y} = \sum m_i y_i$ as well

Example: Four particles of mass 1 kg, 2 kg, 3 kg and 4 kg are attached by light rods of length 50 cm as shown. Find the centre of mass of the structure.

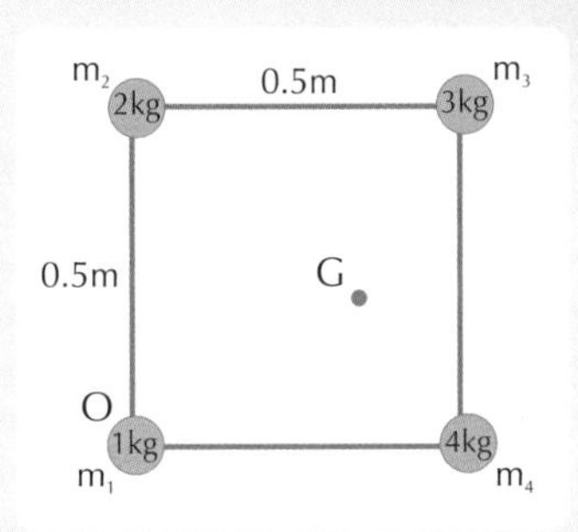

Consider particles as point masses. Take the origin, O, at the 1 kg mass.

Resolving horizontally: $M\bar{x} = m_1x_1 + m_2x_2 + m_3x_3 + m_4x_4$

$10\bar{x} = (1\times0) + (2\times0) + (3\times0.5) + (4\times0.5)$

$\bar{x} = 0.35$ m

Resolving vertically: $M\bar{y} = m_1y_1 + m_2y_2 + m_3y_3 + m_4y_4$

$10\bar{y} = (1\times0) + (2\times0.5) + (3\times0.5) + (4\times0)$

$\bar{y} = 0.25$ m

So the centre of mass is 0.35 m to the right of and 0.25 m above m_1.

Combined Laminas Use Areas Instead of Mass

(because the masses are proportional to the areas)

Example: Two rectangular laminas made of identical material are attached to each other as shown. Take the origin at O. Find the angle OC makes to the vertical when the structure is suspended from corner C.

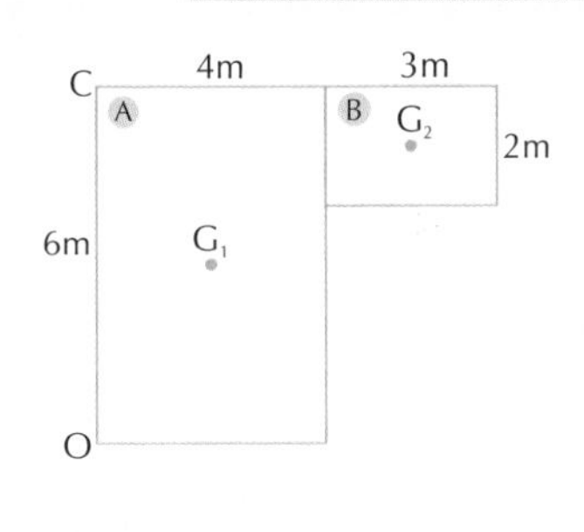

m_1 = Area A $= 6\times4 = 24$ m^2

$x_1 = 2$ m $\quad y_1 = 3$ m ← G_1 lies in the centre of the rectangle.

m_2 = Area B $= 3\times2 = 6$ m^2

$x_2 = 5.5$ m $\quad y_2 = 5$ m

$M\bar{x} = m_1x_1 + m_2x_2$

$30\bar{x} = (24\times2) + (6\times5.5)$

$\bar{x} = 2.7$ m

$M\bar{y} = m_1y_1 + m_2y_2$

$30\bar{x} = (24\times3) + (6\times5)$

So $\bar{y} = 3.4$ m above 0

— which means that it's 2.6 m *below* C.

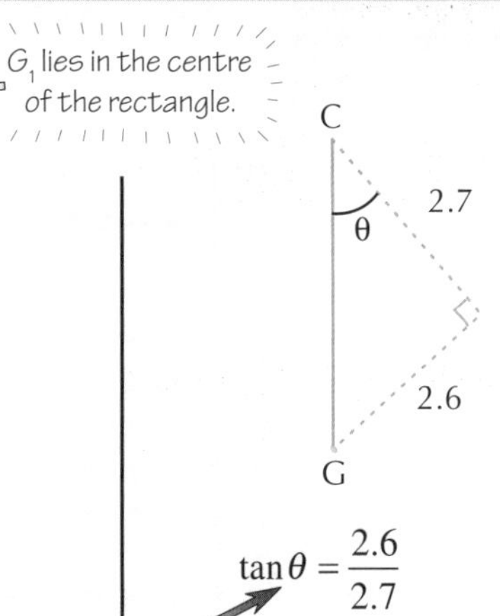

$\tan\theta = \frac{2.6}{2.7}$

So $\theta = 43.9°$

Practice Questions

1) ***Three particles of mass 2 kg, 4 kg and 6 kg are joined by light rods of length 2 m and 3 m as shown. Find the position of the centre of mass relative to O.***

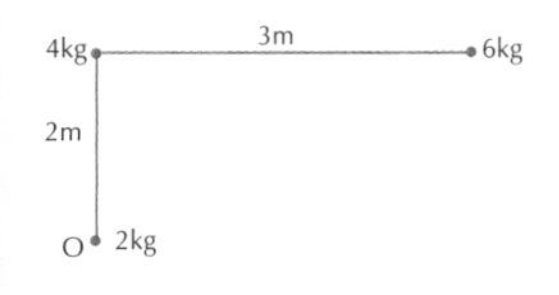

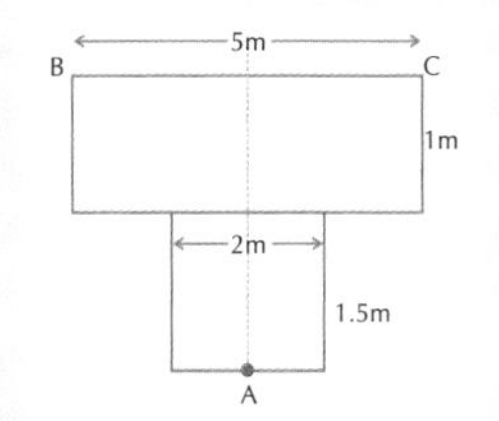

2) ***A symmetrical 'T' shape made of two rectangular laminas of identical material is shown. Find the height above A of the centre of mass of the whole shape and the angle BC makes to the horizontal when the shape is suspended from B.***

3) ***Sample Exam question:***

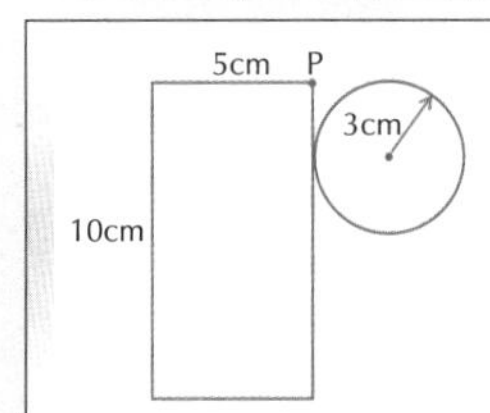

A 10 cm by 5 cm rectangular lamina is attached to a circular lamina (made of the same material) of radius 3 cm at a point 3 cm below P. Find the angle that the 10 cm side of the rectangle makes to the vertical when the structure is suspended from P.

[6 marks]

Phobia #6 — kakorrhiaphobia: fear of failure or defeat... *at least you don't have to worry about that one...*

In the Exam you <u>could</u> cut out shapes and hang them from strings to find centres of mass. But don't bother.

Newton's Laws

That clever chap Isaac Newton established 3 laws involving motion. You need to know all of them.

Newton's First Law	Newton's Second Law	Newton's Third Law
A body will stay at rest or maintain a constant velocity — unless an extra force acts to change that motion.	$F_{net} = ma$ F_{net} (the overall resultant force) is equal to the mass multiplied by the acceleration. Also, F_{net} and a are in the same direction.	For two bodies in contact with each other, the force each applies to the other is equal in magnitude but opposite in direction.

Hint: F_{net} = **ma** is sometimes just written as **F = ma**, but it means the same thing.

Resolve Forces in Perpendicular Directions

Example: A mass of 4 kg at rest on a smooth horizontal plane is acted on by a horizontal force of 5 N. Find the acceleration of the particle and the normal reaction from the plane. Take $g = 9.8 \text{ ms}^{-2}$.

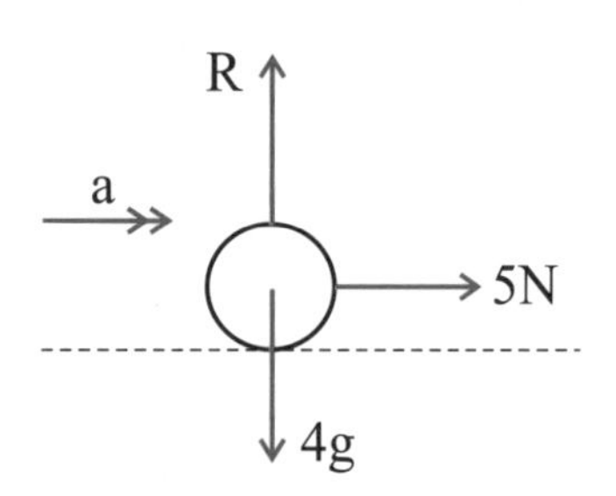

Resolve horizontally:
$F_{net} = ma$ ← Always write $F_{net} = ma$ first.
$5 = 4a$
$a = 1.25 \text{ ms}^{-2}$ to the right

Resolve vertically:
$F_{net} = ma$
$R - 4g = 4 \times 0$
$R = 4g = 39.2 \text{ N}$

Example: A particle of weight 30 N is being accelerated across a smooth plane by a force of 6 N acting at an angle of 25° to the horizontal. Given that the particle starts from rest, find:

a) its speed after 4 seconds,

b) the magnitude of the normal reaction with the plane. What assumptions have you made?

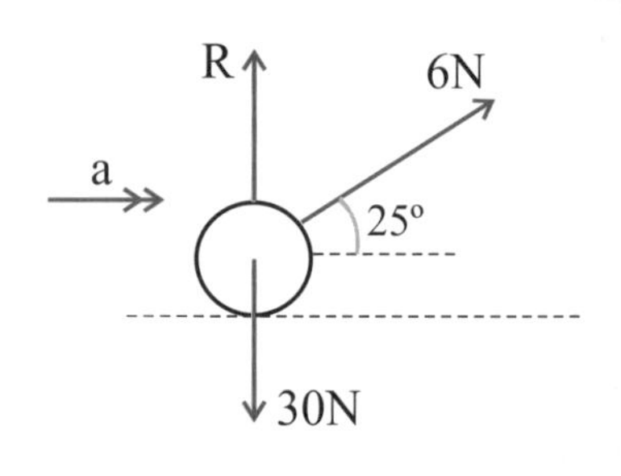

a) Resolve horizontally:
$F_{net} = ma$

$6\cos 25° = \frac{30}{g}a$ so $a = 1.78 \text{ ms}^{-2}$

$v = u + at$

$v = 0 + 1.78 \times 4 = 7.11 \text{ ms}^{-1}$

b) Resolve vertically:
$F_{net} = ma$

$R + 6\sin 25° - 30 = \frac{30}{g} \times 0$

So $R = 30 - 6\sin 25° = 27.5 \text{ N}$ (to 3 s.f.)

Assumptions:
- particle is considered as a point mass,
- there's no air resistance,
- it's a constant acceleration.

Newton's Laws

*You can resolve forces **Parallel** and **Perpendicular** to **Planes***

Example:

A mass of 600 g is propelled up the line of greatest slope of a smooth plane inclined at 30° to the horizontal. If its initial velocity is 3 ms^{-1} find the distance it travels before coming to rest and the magnitude of the normal reaction.

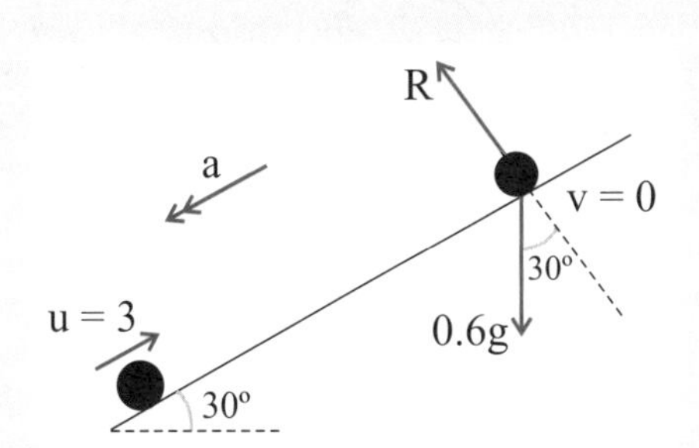

Resolve in ↗ direction:

$F_{net} = ma$

$-0.6g\sin 30 = 0.6a$

$a = -4.9\ ms^{-2}$

$v^2 = u^2 + 2as$

$0 = 3^2 + 2(-4.9)s$

So $s = 0.92$ m

Taking up plane as +ve.

Resolve in ↖ direction:

$F_{net} = ma$

$R - 0.6g\cos 30 = 0.6 \times 0$

So $R = 5.09$ N

Practice Questions

1) A horizontal force of 2 N acts on a 1.5 kg particle initially at rest on a smooth horizontal plane. Find the speed of the particle 3 seconds later.

2) Two forces act on a particle of mass 8 kg which is initially at rest on a smooth horizontal plane. The two forces are (24i+18j) N and (6i+22j) N (with i and j being perpendicular unit vectors in the plane). Find the magnitude and direction of the resulting acceleration of the particle and its displacement after 3 seconds.

3) A horizontal force P acting on a 2 kg mass generates an acceleration of 0.3 ms^{-2}. Given that the mass is in contact with a rough horizontal plane which resists motion with a force of 1 N, find P. Then find the coefficient of friction, μ, to 2 d.p.

4) Sample exam question:

A crane moves a mass of 300 kg, which is modelled as a particle A suspended by two cables AB and AC attached to a movable beam BC. The mass is moved in the direction of the line of the supporting beam BC during which time the cables maintain a constant angle of 40° to the horizontal as shown.

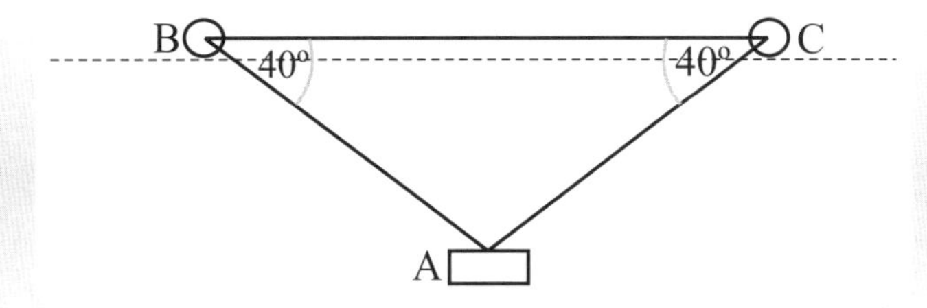

a) The mass is initially moving with constant speed. Find the tension in each cable. [4 marks]

b) The crane then moves the mass with a constant acceleration of 0.4 ms^{-2}. Find the tension in each cable. [6 marks]

c) What modelling assumptions have you made in part b? [2 marks]

Interesting Newton fact: Isaac Newton had a dog called Diamond...

Did you know that Isaac Newton held the same position at Cambridge that Stephen Hawking holds today? And the dog fact about Newton is true — don't ask me how I know such things, just bask in my amazing knowledge of all things trivial.

Friction and Inclined Planes

If you're doing OCR B, you don't need to do calculations involving the coefficient of friction until M2 — you still need to be able to use F_{net} = ma though.

Solving these problems involves careful use of $F_{net} = ma$, $F \leq \mu R$ and the equations of motion.

Use F = ma in *Two Directions* for *Inclined Plane* questions

For inclined slope questions, it's much easier to resolve forces parallel and perpendicular to the plane's surface.

Example: A mass of 3 kg is being pulled up a plane inclined at 20° to the horizontal by a rope parallel to the surface. Given that the mass is accelerating at 0.6 ms⁻² and that the coefficient of friction is 0.4, find the tension in the rope. Take g = 9.8 ms⁻².

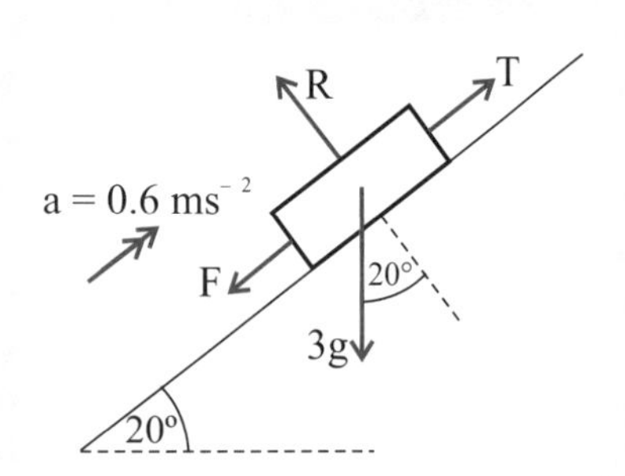

Resolving in ↖ direction: $F_{net} = ma$

$R - 3g\cos 20° = 3 \times 0$

so: $R = 3g\cos 20° = 27.63$ N

The mass is sliding, so $F = \mu R$

$= 0.4 \times 27.62 = 11.05$ N

Resolving in ↗ direction: $F_{net} = ma$

$T - F - 3g\sin 20° = 3 \times 0.6$

$T = 1.8 + 11.05 + 3g\sin 20° = 22.9$ N

Remember that friction always acts in the opposite direction to the motion.

Example: A small body of weight 20 N accelerates from rest and moves a distance of 5 m down a rough plane angled at 15° to the horizontal. Draw a force diagram and find the coefficient of friction between the body and the plane given that the motion takes 6 seconds. Take g = 9.8 ms⁻².

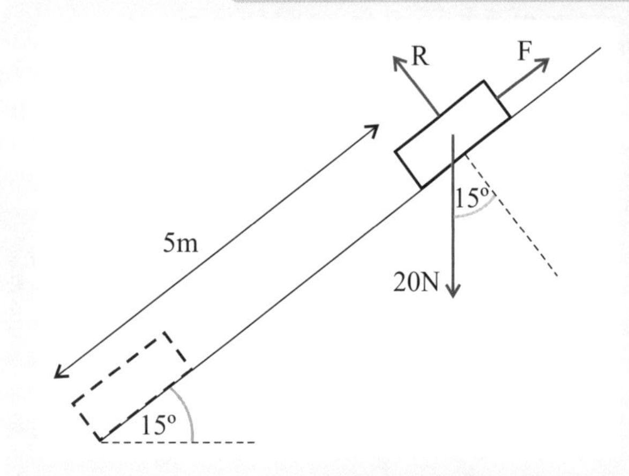

u = 0, s = 5, t = 6, a = ?

Use one of the equations of motion: $s = ut + \frac{1}{2}at^2$

$5 = (0 \times 6) + (\frac{1}{2}a \times 6^2)$ so $\mathbf{a = 0.2778\ ms^{-2}}$

Resolving in ↖ direction: $F_{net} = ma$

$R - 20\cos 15° = \frac{20}{g} \times 0$

So: $R = 20\cos 15° = \mathbf{19.32\ N}$

Resolving in ↙ direction: $F_{net} = ma$

$20\sin 15° - F = \frac{20}{g} \times 0.28$

$\mathbf{F = 4.609\ N}$

It's sliding, so $F = \mu R$

$4.609 = \mu \times 19.32$

$\mu = 0.24$ (to 2 d.p.)

Friction and Inclined Planes

Skip this page if you're doing OCR B.

Friction Opposes Limiting Motion

For a body at rest but on the point of moving down a plane, the friction force is up the plane.
A body about to move up a plane is opposed by friction down the plane.

Example: A 4 kg box is placed on a 30° plane where $\mu = 0.4$. A force Q maintains equilibrium by acting up the plane parallel to the line of greatest slope. Find Q if the box is on the point of sliding a) up the plane, b) down the plane.

a)

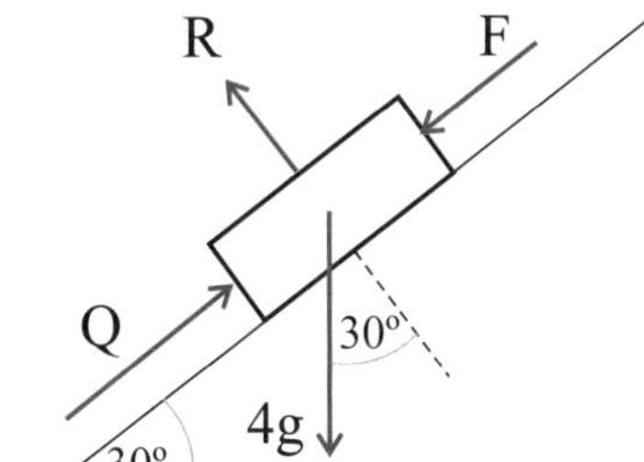

$F_{net} = ma$

Resolving in ↖ direction:

$R - 4g\cos30° = 0$

$R = 4g\cos30°$

$F = \mu R$

$= 0.4 \times 4g\cos30$

$= 1.6g\cos30$

Resolving in ↗ direction:

$Q - 4g\sin30 - F = 4 \times 0$

$Q = 4g\sin30 + 1.6g\cos30$

$= 33.2$ N

b)

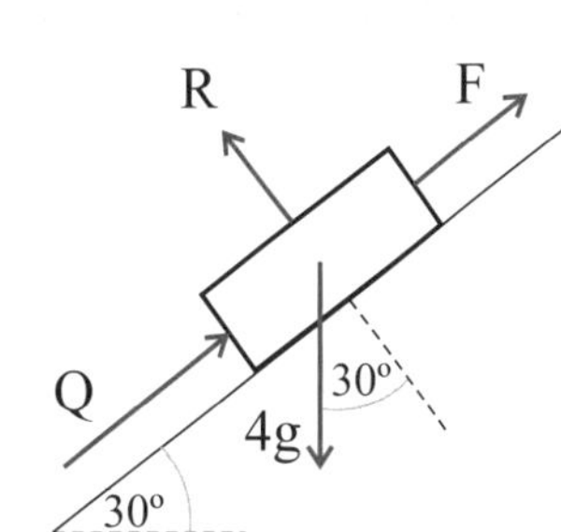

Resolving in ↖ direction:

$R = 4g\cos30$

$F = 1.6g\cos30$

Resolving in ↗ direction:

$Q - 4g\sin30 + F = 4 \times 0$

$Q = 4g\sin30 - 1.6g\cos30$

$Q = 6.02$ N

So for equilibrium $6.02 \text{ N} \leq Q \leq 33.2 \text{ N}$

Practice Questions

1) ***A brick of mass 1.2 kg is sliding down a rough plane which is inclined at 25° to the horizontal. Given that its acceleration is 0.3 ms⁻², find the coefficient of friction between the brick and the plane. What assumptions have you made?***

2) ***An army recruit of weight 600 N steps off a tower and accelerates down a "death slide" wire as shown. The recruit hangs from a light rope held between her hands and looped over the wire. The coefficient of friction between the rope and wire is 0.5. Given that the wire is 20 m long and makes an angle of 30° to the horizontal throughout its length, find how fast the recruit is travelling when she reaches the end of the wire.***

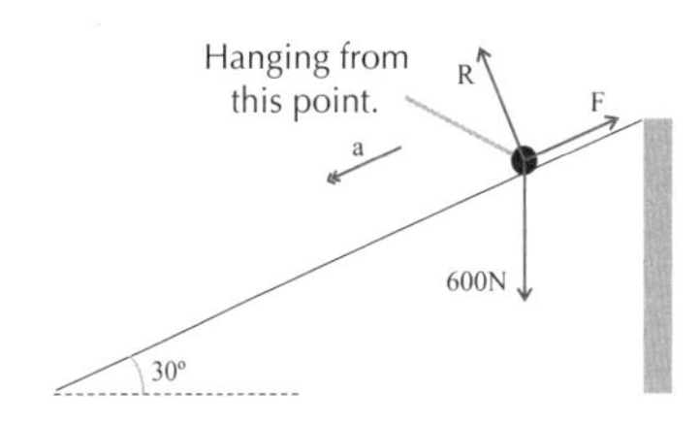

3) ***Sample exam question:***

A horizontal force of 8 N just stops a mass of 7 kg from sliding down a plane inclined at 15° to the horizontal as shown.

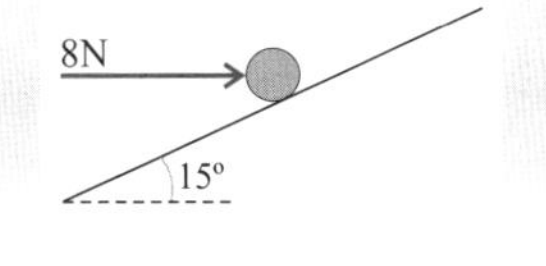

a) Calculate the coefficient of friction between the mass and the plane to 2 d.p. [6 marks]

b) The 8 N force is now removed. Find how long the mass takes to slide a distance of 3 m down the line of greatest slope. [7 marks]

Inclined planes — nothing to do with suggestible Boeing 737s...

The main thing to remember is that you can choose to resolve in any two directions as long as they're perpendicular. It makes sense to choose the directions that involve doing as little work as possible. Obviously.

Constant Acceleration and Path Equations

Skip this page if you're doing OCR A.

You can use column vectors and **i,j** notation when working with equations of motion. Thought you'd be pleased.

*All vectors have **Magnitude** and **Direction***

Example:

A particle P moves on a smooth horizontal plane with constant acceleration. Initially the particle has a velocity of 2**i** ms^{-1} and starts from an origin O. 5 seconds later its velocity is 10**j** ms^{-1}, where **i** and **j** are perpendicular unit vectors in the plane.

Find the magnitude and direction of the acceleration, plus the position of P, 5 seconds after motion begins.

First list the variables for equations of motion:

$$\mathbf{u} = \begin{pmatrix} 2 \\ 0 \end{pmatrix}; \quad \mathbf{v} = \begin{pmatrix} 0 \\ 10 \end{pmatrix}; \quad t = 5; \quad \mathbf{a} = ?$$

In this question, letters in bold (e.g. **a**) refer to vectors. Letters not in bold (e.g. a) mean the magnitude of that vector.

You need to use an equation containing **u**, **v**, t and **a**:

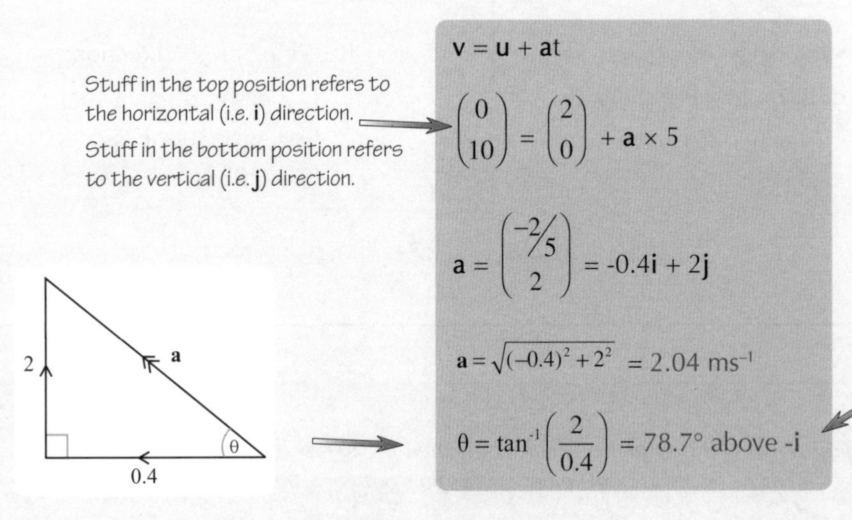

$$\mathbf{v} = \mathbf{u} + \mathbf{a}t$$

$$\begin{pmatrix} 0 \\ 10 \end{pmatrix} = \begin{pmatrix} 2 \\ 0 \end{pmatrix} + \mathbf{a} \times 5$$

$$\mathbf{a} = \begin{pmatrix} -2/5 \\ 2 \end{pmatrix} = -0.4\mathbf{i} + 2\mathbf{j}$$

$$a = \sqrt{(-0.4)^2 + 2^2} = 2.04 \text{ ms}^{-1}$$

$$\theta = \tan^{-1}\left(\frac{2}{0.4}\right) = 78.7° \text{ above } -\mathbf{i}$$

Don't just write the angle — you need to say where it is in relation to **i** or **j**.

So you've got the direction and magnitude of the acceleration sorted out — now you need the position of P after 5 seconds of motion.

Play it safe and make another list of all the variables you know, including the one you've just worked out:

$$\mathbf{u} = \begin{pmatrix} 2 \\ 0 \end{pmatrix}; \quad \mathbf{v} = \begin{pmatrix} 0 \\ 10 \end{pmatrix}; \quad t = 5; \quad \mathbf{a} = \begin{pmatrix} -2/5 \\ 2 \end{pmatrix}$$

You need to know **s**, so choose an equation containing **s**, t, **a** and either **u** or **v**:

$$\mathbf{s} = \mathbf{u}t + \frac{1}{2}\mathbf{a}t^2$$

$$\begin{pmatrix} x \\ y \end{pmatrix} = \begin{pmatrix} 2 \\ 0 \end{pmatrix} \times 5 + \frac{1}{2}\begin{pmatrix} -0.4 \\ 2 \end{pmatrix} \times 5^2 = \begin{pmatrix} 5 \\ 25 \end{pmatrix}$$

i.e. when t = 5, $\overrightarrow{OP} = 5\mathbf{i} + 25\mathbf{j}$

Constant Acceleration and Path Equations

Path Equations give the Position of a Particle in terms of Time

Path equations always contain t, so that you can work out the x and y coordinates of the particle at a particular time. If you eliminate t, you can get values for the coordinates of the particle.

Example:

The position vector, **r**, of a particle is given by **r** = 5t**i** – 5t²**j** relative to perpendicular unit vectors **i** and **j**, based on an origin 0.
Assuming motion begins at 0, sketch the path of the particle during the first 2 seconds of motion and find its path equation for the general position (x,y).

First things first... Start by sketching a quick table showing the x and y values at t = 0, 1 and 2.

	t	0	1	2
x	5t	0	5	10
y	$-5t^2$	0	-5	-20

Then you can plot the points as a graph:

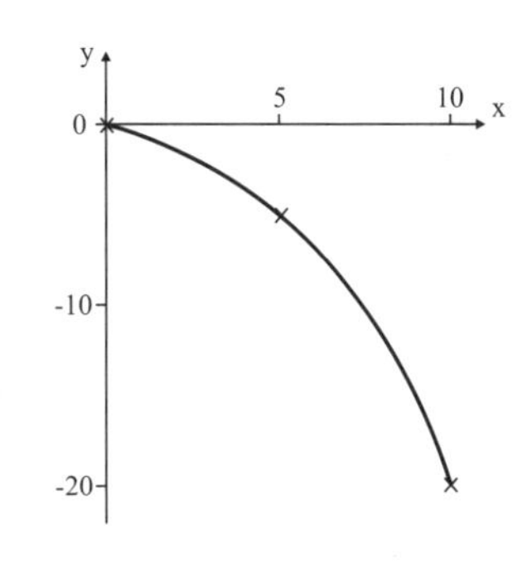

You know for any time t that x = 5t and $y = -5t^2$.
Just rearrange to eliminate t, to give you an equation for the general position (x,y).

$x = 5t$ so $t = \frac{x}{5}$

Plug this into y: $y = -5t^2 = -5\left(\frac{x}{5}\right)^2 = -\frac{1}{5}x^2$

Practice Questions

1) ***A particle moves with constant acceleration away from an origin O. Initially it is moving due south with a speed of 6 ms⁻¹. 20 seconds later it is moving due east with a speed of 8 ms⁻¹. Unit vectors i and j are in directions east and north respectively.***
   a) ***Find the acceleration of the particle in i, j form and find its magnitude and direction.***
   b) ***Find the position of the particle after 5 seconds.***

2) ***Sample exam question:***

A small aircraft flies low over a field. As it pulls sharply up, its path relative to an origin 0 at the end of the field is modelled by the path equation $\mathbf{r} = (15t + 10)\mathbf{i} + (15\sqrt{3}\,t - 5t^2)\mathbf{j}$ for $t > 1$ where **i** and **j** are unit vectors relative to 0 in horizontal and vertical directions respectively. Distances are in metres.

a) How far is the aircraft from the origin when t = 3? [2 marks]

b) Sketch the path of the aircraft's flight for $1 \le t \le 3$. [4 marks]

c) Show that the path equation for the aircraft's motion can be approximated to $y = 2.2x - 20 - 0.02x^2$. [4 marks]

d) Find the expressions for the aircraft's velocity and acceleration at time, t. [3 marks]

e) Suggest why the model is not appropriate for all values of t for $t > 1$. [1 marks]

I promise I'm not leading you up the garden path equation...

Even if you're working with column vectors, these questions are all OK — just plug the values into the equation of motion that gives you the values you want. And path equations are good for working out exactly where that particle's got to.

Connected Particles

Like Laurel goes with Hardy and Posh goes with Becks, some particles are destined to be together...

Connected Particles act like One Mass

Particles connected together have the same speeds and accelerations as each other, unless the connection fails. Train carriages moving together have the same acceleration.

Example: A 30 tonne locomotive engine is pulling a single 10 tonne carriage as shown. They are accelerating at 0.3 ms^{-2} due to the force P generated by the engine. It's assumed that there are no forces resistant to motion. Find P and the tension in the coupling.

Here's the pretty picture:

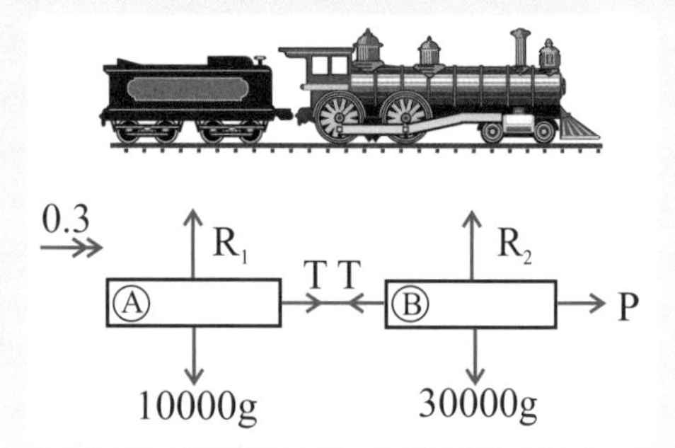

And here's the ugly maths:

For A: $F_{net} = ma$
$T = 10\,000 \times 0.3$
$T = 3000$ N

For B: $F_{net} = ma$
$P - T = 30\,000 \times 0.3$
$P = 12000$ N

Pulleys (and 'Pegs') are always Smooth

In M1 questions, you can always assume that the tension in a string will be the same either side of a smooth pulley.

Example: Masses of 3 kg and 5 kg are connected by an inextensible string and hang vertically either side of a smooth pulley. They are released from rest. Find their acceleration and the time it takes for each to move 40 cm. State any assumptions made in your model.

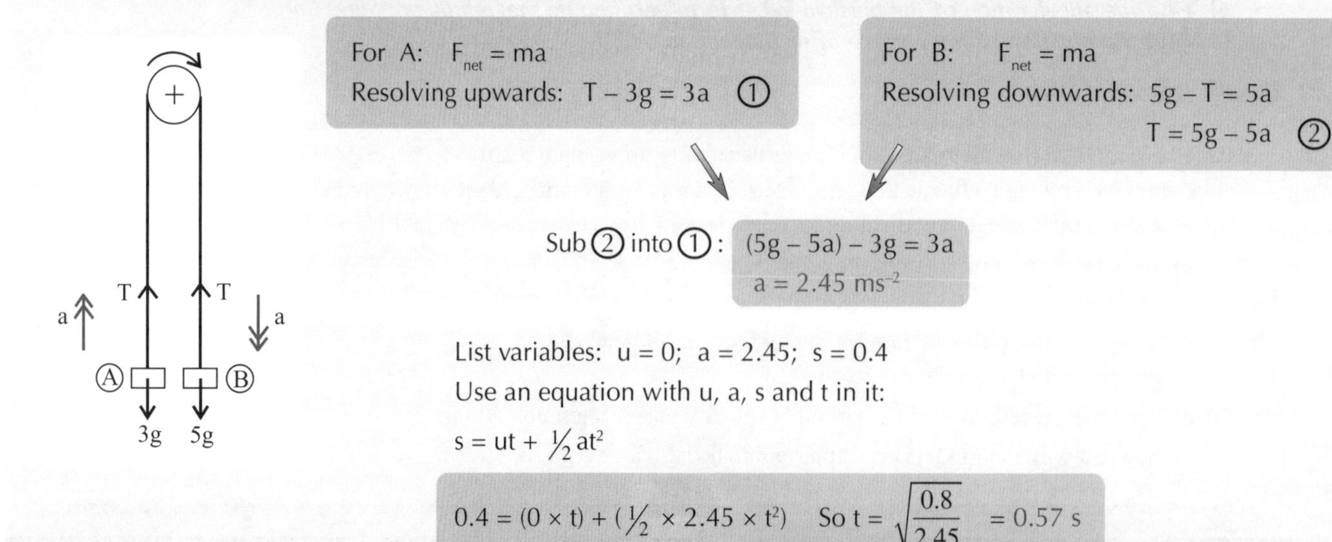

For A: $F_{net} = ma$
Resolving upwards: $T - 3g = 3a$ ①

For B: $F_{net} = ma$
Resolving downwards: $5g - T = 5a$
$T = 5g - 5a$ ②

Sub ② into ①: $(5g - 5a) - 3g = 3a$
$a = 2.45$ ms^{-2}

List variables: $u = 0$; $a = 2.45$; $s = 0.4$
Use an equation with u, a, s and t in it:
$s = ut + \frac{1}{2}at^2$

$0.4 = (0 \times t) + (\frac{1}{2} \times 2.45 \times t^2)$ So $t = \sqrt{\frac{0.8}{2.45}} = 0.57$ s

Assumptions: The 3 kg mass does not hit the pulley; there's no air resistance; the string is 'light' and doesn't break.

Connected Particles

Use F = ma in the Direction Each Particle Moves

Example: A mass of 3 kg is placed on a smooth horizontal table. A light inextensible string connects it over a smooth peg to a 5 kg mass which hangs vertically as shown. Find the tension in the string if the system is released from rest. Take g = 9.8 ms^{-2}.

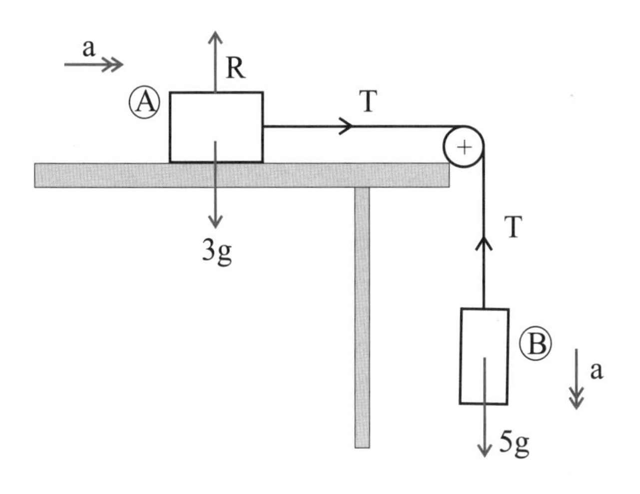

For A:
Resolve horizontally: $F_{net} = ma$

$T = 3a$

$a = \frac{T}{3}$ ①

For B:
Resolve vertically: $F_{net} = ma$

$5g - T = 5a$ ②

Sub ① into ②: $5g - T = 5 \times \frac{T}{3}$

So $\frac{8}{3}T = 5g$

$T = 18.4$ N (to 3 s.f.)

Practice Questions

1) ***A 2 tonne tractor experiences a resistance force of 1000 N whilst pulling a 1 tonne trailer along a horizontal road. If the tractor engine provides a forward force of 1500 N find the resistance force acting on the trailer, and the tension in the coupling between tractor and trailer, if they are moving with constant speed.***
2) ***Two particles are connected by a light inextensible string, and hang in a vertical plane either side of a smooth pulley. When released from rest the particles accelerate at 1.2 ms^{-2}. If the heavier mass is 4 kg, find the weight of the other.***
3) ***Sample exam question:***

A car of mass 1500 kg is pulling a caravan of mass 500 kg.
They experience resistance forces totalling 1000 N and 200 N respectively.
The forward force generated by the car's engine is 2500 N. The coupling between the two does not break.

a) Find the acceleration of the car and caravan. [3 marks]

b) Find the tension in the coupling. [3 marks]

Useful if you're hanging over a Batman-style killer crocodile pit...

It makes things a lot easier when you know that connected particles act like one mass, and that in M1 pulleys can always be treated as smooth. Those examiners occasionally do try and make your life easier, honestly.

Connected Particles

If you're doing OCR B, you don't need to use the coefficient of friction in calculations until M2.

More complicated pulley and peg questions have friction and inclined planes for you to enjoy too.

*Remember to use $F \le \mu R$ on **Rough Planes***

For a particle on a plane, don't forget to resolve the forces in two directions.

Example:

The peg system of the example on p31 is set up again. However, this time a friction force, F, acts on the 3 kg mass due to the table top now being rough, with coefficient of friction $\mu = 0.5$. Find the new tension in the string when the particles are released from rest.

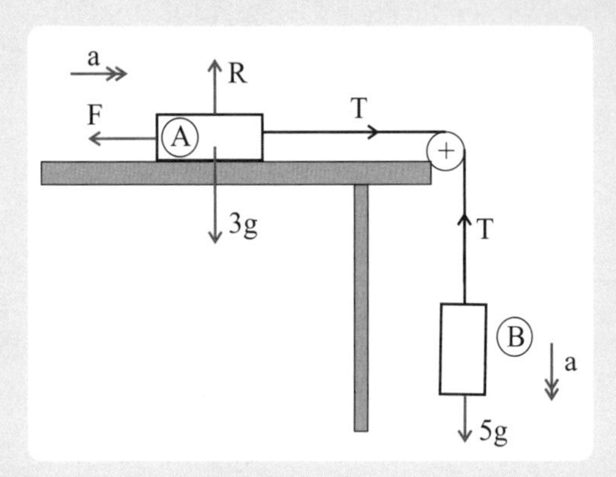

For B: Resolving vertically: $5g - T = 5a$ ①

For A: Resolving horizontally: $F_{net} = ma$

$T - F = 3a$ ②

Resolving vertically: $R - 3g = 0$

$\mathbf{R = 3g}$

The particles are moving, so $F = \mu R = 0.5 \times 3g$

$= \mathbf{14.7\ N}$

Sub this into ②: $T - 14.7 = 3a$

$$\mathbf{a = \frac{1}{3}(T - 14.7)}$$

Sub this into ①: $5g - T = 5 \times \frac{1}{3}(T - 14.7)$

$8T = 147 + 73.5$

$T = 27.6\ N$

***Rough Inclined Plane** questions need **Really Good** force diagrams*

You know the routine... resolve forces parallel and perpendicular to the plane... yawn.

Example:

A 3 kg mass is held in equilibrium on a rough ($\mu = 0.4$) plane inclined at 30° to the horizontal. It is attached by a piece of light, inextensible string to a mass M kg hanging vertically beneath a smooth pulley, as shown. Find M if the 3 kg mass is on the point of sliding up the plane.

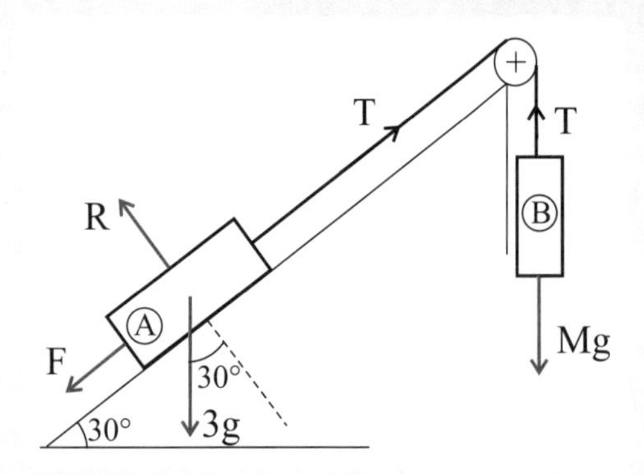

For B: Resolving vertically: $F_{net} = ma$

$Mg - T = M \times 0$

$\mathbf{T = Mg}$

For A: Resolving in ↖ direction: $F_{net} = ma$

$R - 3g\cos30 = 3 \times 0$

$\mathbf{R = 3g\cos30}$

It's limiting friction, so: $F = \mu R$

$F = 0.4 \times 3g\cos30$

$= \mathbf{10.18\ N}$

For A: Resolving in ↗ direction: $F_{net} = ma$

$T - F - 3g\sin30 = 3 \times 0$

$Mg - 10.18 - 3g\sin30 = 0$

$M = 2.54$ kg (to 3 s.f.)

Connected Particles

Example:

Particles A and B of mass 4 kg and 5 kg are connected by a light inextensible string over a smooth pulley as shown. A force of Q acts on A at an angle of 25° to the rough ($\mu = 0.6$) horizontal plane. Find the range of values of the magnitude of Q if equilibrium is to be maintained.

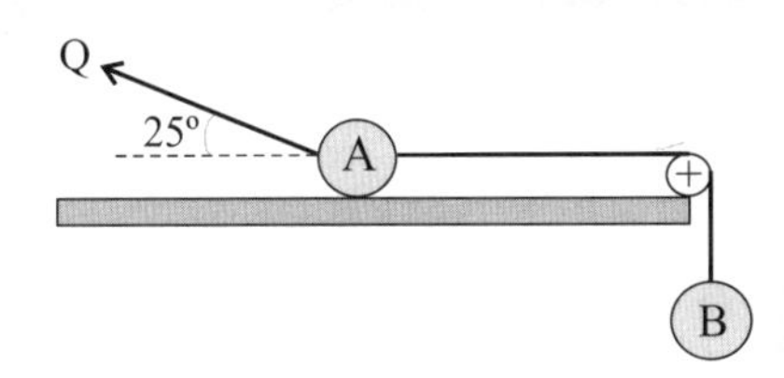

You need to work out Q if A is about to move left or right.

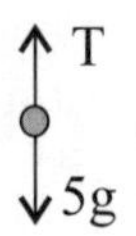

Mass B is the same whichever way the frictional force F acts:

Resolving vertically: $T = 5g$ ①

i) Work out Q if A is about to go right:

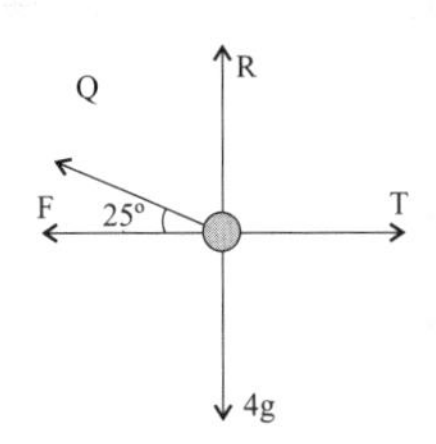

Resolving vertically: $R = 4g - Q\sin 25°$

$F = \mu R$

$= 0.6(4g - Q\sin 25°)$ ②

Resolving horizontally: $T - Q\cos 25° - F = 0$

Using equations 1 and 2. $5g - Q\cos 25° - 0.6(4g - Q\sin 25°) = 0$

So $Q = 39.0$ N (to 3 s.f.)

ii) Work out Q if A is about to go left:

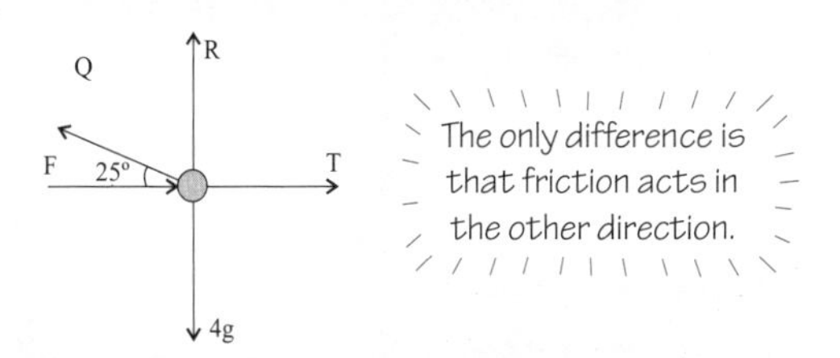

Resolving vertically: $R = 4g - Q\sin 25°$

$F = \mu R$

$= 0.6(4g - Q\sin 25°)$

Resolving horizontally: $Q\cos 25° - F - T = 0$

$Q\cos 25° - 0.6(4g - Q\sin 25°) - 5g = 0$

So $Q = 62.5$ N (to 3 s.f.)

Practice Questions

1) ***Particles of mass 3 kg and 4 kg are connected by a light, inextensible string passing over a smooth pulley as shown. The 3 kg mass is on a smooth slope angled at 40° to the horizontal. Find the acceleration of the system if released from rest, and find the tension in the string. What minimum force acting on the 3 kg mass parallel to the plane would be needed to maintain equilibrium?***

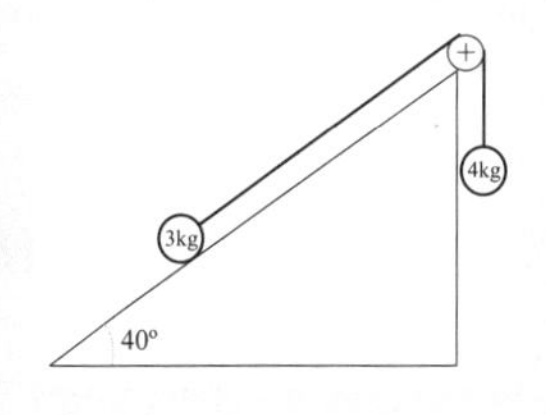

2) ***Sample exam question:***

Two particles P and Q of masses 1 kg and m kg respectively are linked by a light string passing over a smooth pulley as shown. Particle P is on a rough slope inclined at 20° to the horizontal, where the coefficient of friction between P and the plane is 0.1.

a) Given that P is about to slide down the plane, find the mass of Q. [5 marks]

b) Describe the motion of the system if the mass of Q is 1kg. [5 marks]

P Q 20°

Connected particles — together forever... *isn't it beautiful?*

The key word here is rough. If a question mentions the surface is rough, then cogs should whirr and the word 'friction' should pop into your head. Take your time with force diagrams of rough inclined planes — I had a friend who rushed into drawing a diagram, and he ended up with a broken arm. But that was years later, now that I come to think of it.

Momentum

Skip these two pages if you're doing OCR B.

Momentum is a measure of how much strength a moving object has, due to its mass and velocity.

Momentum has Magnitude and Direction

Total momentum before a collision equals total momentum after a collision. This idea is called "Conservation of Momentum".

Because it's a vector, the sign of the velocity in momentum is important.

Momentum = Mass × Velocity

Example: Particles A and B, each of mass 5 kg, move in a straight line with velocities 6 ms^{-1} and 2 ms^{-1} respectively. After collision mass A continues in the same direction with velocity 4.2 ms^{-1}. Find the velocity of B after impact.

Before

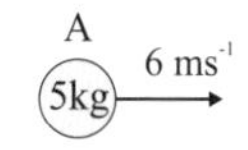

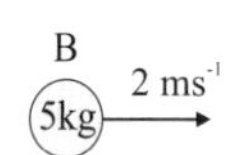

After

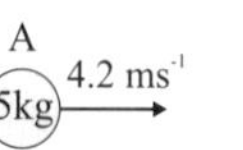

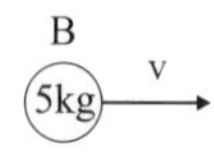

Draw 'before' and 'after' diagrams to help you see what's going on.

Momentum A + Momentum B = Momentum A + Momentum B

$$(5 \times 6) + (5 \times 2) = (5 \times 4.2) + (5 \times v)$$

$$40 = 21 + 5v$$

So $v = 3.8$ ms^{-1} in the same direction as before

Example: Particles A and B of mass 6 kg and 3 kg are moving towards each other at speeds of 2 ms^{-1} and 1 ms^{-1} respectively. Given that B rebounds with speed 3 ms^{-1} in the opposite direction to its initial velocity, find the velocity of A after the collision.

Before

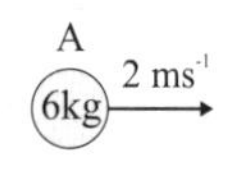

B: 3kg, 1 ms^{-1} (towards A)

After

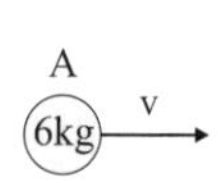

B: 3kg, 3 ms^{-1}

$$(6 \times 2) + (3 \times -1) = (6 \times v) + (3 \times 3)$$

$$9 = 6v + 9$$

$$v = 0$$

Masses Joined Together have the Same Velocity

Particles that stick together after impact are said to "coalesce". After that you can treat them as just one object.

Example: Two particles of mass 40 g and M kg move towards each other with speeds of 6 ms^{-1} and 3 ms^{-1} respectively. Given that the particles coalesce after impact and move with a speed of 2 ms^{-1} in the same direction as that of the 40 g particle's initial velocity, find M.

Before

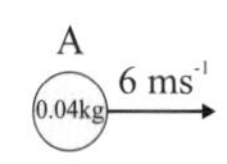

B: M, 3 ms^{-1} (towards A)

After

(M + 0.04) kg, 2 ms^{-1}

$$(0.04 \times 6) + (M \times -3) = [(M + 0.04) \times 2]$$

$$0.24 - 3M = 2M + 0.08$$

$$5M = 0.16$$

$$M = 0.032 \text{ kg} = 32 \text{ g}$$

Don't forget to convert all masses to the same units.

Momentum

Example:

A lump of ice of mass 0.1 kg is thrown across the surface of a frozen lake with speed 4 ms^{-1}. It collides with a stationary stone of mass 0.3 kg. The lump of ice and the stone then move in opposite directions to each other with the same speeds. Find their speed.

Before

A (0.1kg) → 4 ms^{-1} B (0.3kg) 0 ms^{-1}

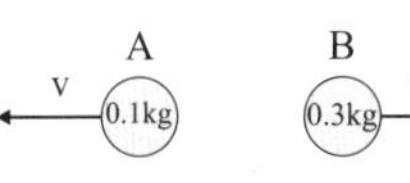

$$(0.1 \times 4) + (0.3 \times 0) = (0.1 \times -v) + 0.3v$$

$$0.4 = -0.1v + 0.3v$$

$$v = 2 \text{ ms}^{-1}$$

Practice Questions

Each diagram represents the motion of two particles moving in a straight line. Find the missing mass or velocity. (All masses are in kg and all velocities are in ms^{-1}.)

	Before	***After***
1)	(5) 3 → (4) 1 →	(5) 2 → (4) v →
2)	(5) 3 → (4) 1 →	(9) v →
3)	(5) 3 → ← 2 (4)	← v (5) (4) 3 →
4)	(m) 6 → (8) 2 →	(m) 2 → (8) 4 →

5) Sample exam question:

Two particles of mass 0.8 kg and 1.2 kg are travelling in the same direction along a straight line with speeds of 4 ms^{-1} and 2 ms^{-1} respectively. After collision the 0.8 kg mass has a velocity of 2.5 ms^{-1} in the same direction. The 1.2 kg mass then continues with its new velocity until it collides with a mass m kg travelling with a speed of 4 ms^{-1} in the opposite direction to it. Both particles are brought to rest by this collision. Find the mass m. [4 marks]

Ever heard of Hercules?

Well, he carried out 12 tasks, including: killing the Nemean lion, capturing the Erymanthian boar, acquiring the golden apples of the Hesperides, killing the Stymphalian birds, cleaning the stables of Augeas and capturing the girdle of Hippolyte. Nothing to do with momentum, but next time you're feeling sorry for yourself for doing M1, think on.

Impulse

Skip these two pages if you're doing OCR A, OCR B, AQA A or AQA B.

An impulse changes the momentum of a particle in the direction of motion.

*Impulse is **Change in Momentum***

To work out the impulse that's acted on an object, just subtract the object's initial momentum from its final momentum. Impulse is measured in newton seconds (Ns).

Impulse = mv – mu

Example:

A body of mass 500 g is travelling in a straight line. Find the magnitude of the impulse needed to increase its speed from 2 ms^{-1} to 5 ms^{-1}.

Impulse = Change in momentum
= mv – mu
= (0.5 × 5) – (0.5 × 2)
= 1.5 Ns

Momentum = mass × velocity

Example:

A 20 g ball is dropped 1 m onto the ground. Immediately after rebounding the ball has a speed of 2 ms^{-1}. Find the impulse given to the ball by the ground. How high does the ball rebound?

First you need to work out the ball's speed as it reaches the ground:

List the variables you're given:

u = 0 ← *The ball was dropped, so it started from u = 0.*
s = 1
a = 9.8 ← *Acceleration due to gravity.*
v = ?

Choose an equation containing u, s, a and v:

$v^2 = u^2 + 2as$
$v^2 = 0^2 + (2 \times 9.8 \times 1)$
$v = 4.43\ ms^{-1}$ ← *The sign is really important. Make sure that down is positive in this part of the question.*

Now work out the impulse as the ball hits the ground and rebounds:

u = 4.43 ms^{-1} ↓ v = 2 ms^{-1} ↑ I ↑

Impulse = mv – mu
= (0.02 × 2) – (0.02 × –4.43)
= 0.129 Ns (to 3 s.f.)

Finally you need to use a new equation of motion to find s (the greatest height the ball reaches).

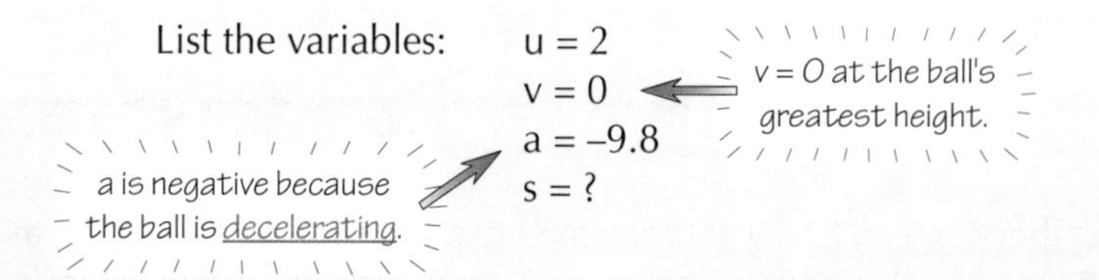

$v^2 = u^2 + 2as$
$0^2 = 2^2 + (2 \times -9.8 \times s)$
s = 0.204 m (to 3 s.f.)

Impulse

Impulses always **Balance** *in* **Collisions**

During impact between particles A and B, the impulse that A gives to B is the same as the impulse that B gives to A, but in the opposite direction.

Example:

A mass of 2 kg moving at 2 ms^{-1} collides with a mass of 3 kg which is moving in the same direction at 1 ms^{-1}. The 2 kg mass continues to move in the same direction at 1 ms^{-1} after impact. Find the impulse given by the 2 kg mass to the other particle.

Using "conservation of momentum":

$$(2 \times 2) + (3 \times 1) = (2 \times 1) + 3v$$

$$\text{So} \quad v = 1\frac{2}{3}$$

Before

A (2kg) 2 ms^{-1} → B (3kg) 1 ms^{-1} →

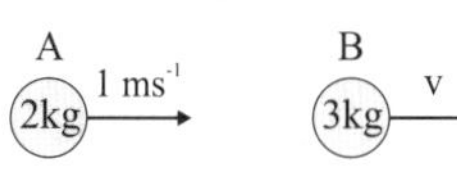

Impulse (on B) = mv – mu (for B)

$$= (3 \times 1\frac{2}{3}) - (3 \times 1)$$

$$= \mathbf{2\ Ns}$$

The impulse B gives to A is $(2 \times 1) - (2 \times 2) =$ **–2 Ns**. Aside from the different direction, you can see it's the same — so you didn't actually need to find v for this question.

Impulse *is linked to* **Force** *too*

Impulse is also related to the force needed to change the momentum and the time it takes.

Impulse = Force × Time

Example:

A 0.9 tonne car increases its speed from 30 kmh^{-1} to 40 kmh^{-1}. Given that the maximum additional forward force the car's engine can produce is 1 kN, find the shortest time it will take to achieve this change in speed.

Impulse = mv – mu

$$= (900 \times \frac{40\,000}{3600}) - (900 \times \frac{30\,000}{3600})$$

$$= 2500 \text{ Ns}$$

Now use Impulse = Force × Time:

$$2500 = 1000 \times t$$

$$t = 2.5 \text{ s}$$

Practice Questions

1) ***An impulse of 2 Ns acts against a ball of mass 300 g moving with a velocity of 5 ms^{-1}. Find the ball's new velocity.***

2) ***A particle of mass 450 g is dropped 2 m onto a floor. It rebounds to two thirds of its original height. Find the impulse given to the ball by the ground.***

3) ***Sample exam question:***

A coal wagon of mass 4 tonnes is rolling along a straight rail track at 2.5 ms^{-1}. It collides with a stationary wagon of mass 1 tonne. During the collision the wagons become coupled and move together along the track.

a) Find their speed after collision. [2 marks]

b) Find the impulse given to the more massive wagon. [2 marks]

c) State two assumptions made in your model. [2 marks]

Still feeling dynamic? Just wait for the projectiles section...

Impulse is change in momentum — remember that and you'll be laughing. Anyway, I know all about impulse. Those orange nylon flares looked great in the shop window, but I really should have tried them on before I bought them.

Projectiles

Skip these two pages if you're doing Edexcel or OCR A.

A 'projectile' is just any old object that's been lobbed through the air. When you're doing projectile questions you'll have to model the motion of particles in two dimensions whilst ignoring air resistance.

Split Motion into **Horizontal** and **Vertical** Components

It's time that connects the two directions.
Remember that the only acceleration is due to gravity — so horizontal acceleration is zero.

Example: A stone is thrown horizontally with speed 10 ms^{-1} from a height of 2 m above the horizontal ground. Find the time taken for the stone to hit the ground and the horizontal distance travelled before impact. Find also the stone's velocity after 0.5 s.

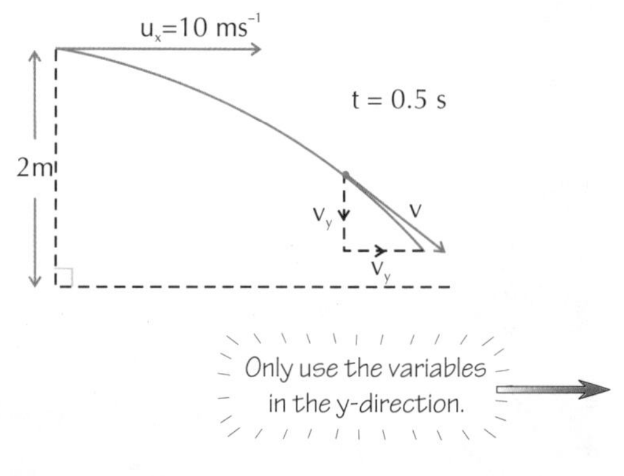

Only use the variables in the y-direction.

Resolving vertically (take down as +ve):

u = 0 s = 2
a = 9.8 t = ?

$s = ut + \frac{1}{2}at^2$

$2 = 0 \times t + \frac{1}{2} \times 9.8 \times t^2$

t = 0.64 s (to 2 s.f.)
i.e. the stone lands after 0.64 seconds

Resolving horizontally (take right as +ve):

u = 10 s = ?
a = 0 t = 0.64

$s = ut + \frac{1}{2}at^2$

$= 10 \times 0.64 + \frac{1}{2} \times 0 \times 0.64^2$

= 6.4 m

i.e. the stone has gone 6.4 m horizontally when it lands.

Now find the velocity after 0.5 s — again, keep the horizontal and vertical bits separate.

v = u + at
$v_y = 0 + 9.8 \times 0.5$
= 4.9 ms^{-1}

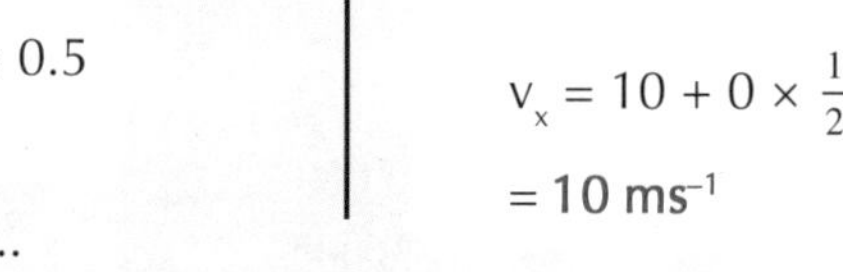

v = u + at
$v_x = 10 + 0 \times \frac{1}{2}$
= 10 ms^{-1}

Now you can find the speed and direction if you want to...

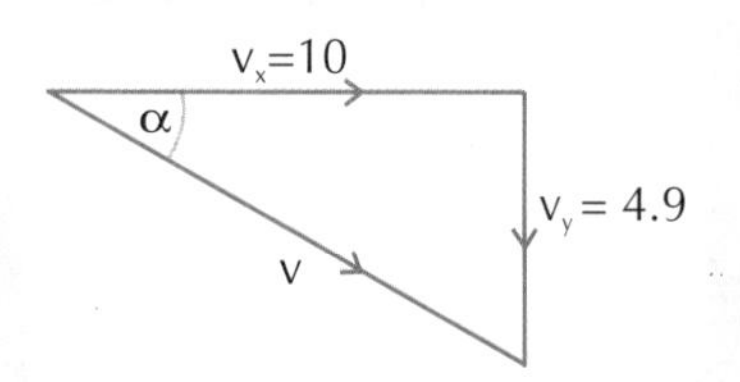

$v = \sqrt{4.9^2 + 10^2} = 11.1 \text{ ms}^{-1}$

$\tan\theta = \frac{4.9}{10}$

So $\theta = 26.1°$ below horizontal

Split **Velocity of Projection** into **Two Components** too

A particle projected with a speed U at an angle α to the horizontal has two components of initial velocity.

Here's the same information in a diagram:

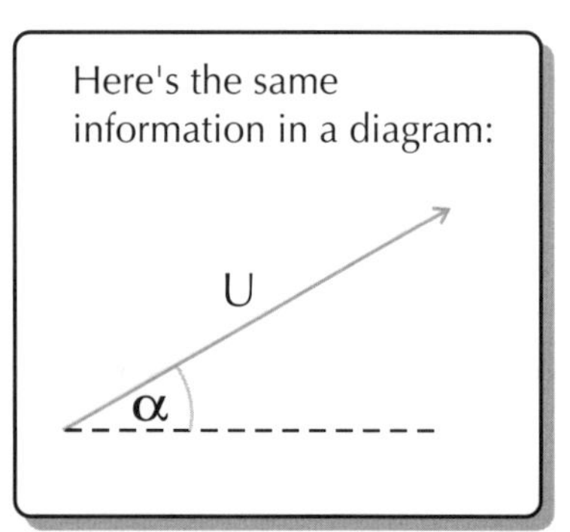

Split the velocity into its x and y components:

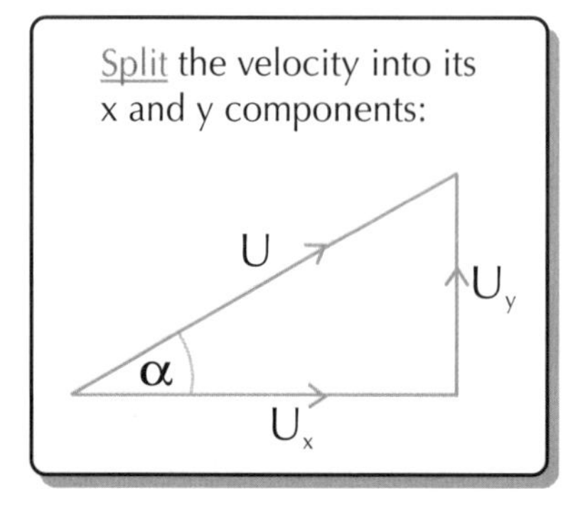

Finally, work out the values of the components using trigonometry:

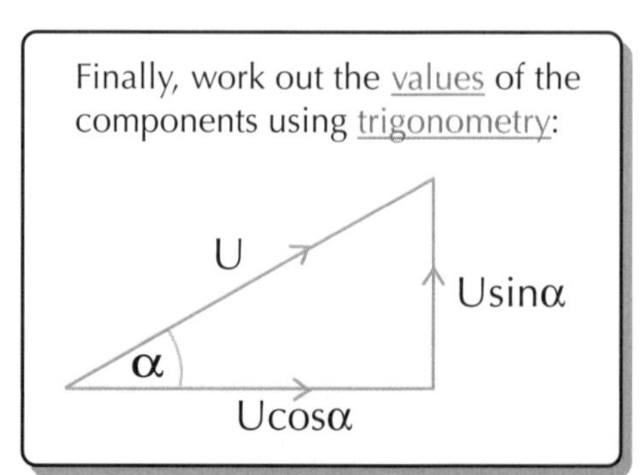

Projectiles

Example:

A cricket ball is projected with a speed of 30 ms^{-1} at an angle of 25° to the horizontal.

a) Find the maximum height it reaches (h) and the horizontal distance travelled (r). Assume that the ground is horizontal and that the ball is struck at ground level.

b) If instead the ball is struck 1.5 m above the ground, find the new maximum height and new horizontal distance travelled.

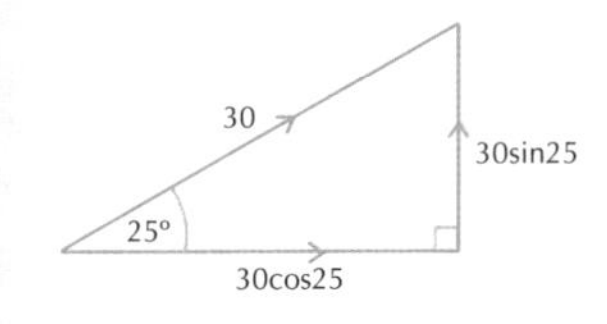

a) **Resolving vertically** (take up as +ve):

u = 30sin25°
v = 0
a = –9.8
s = h

$$v^2 = u^2 + 2as$$
$$0 = (30\sin 25°)^2 + 2(-9.8 \times h)$$
$$h = 8.2 \text{ m}$$

$$v = u + at$$
$$0 = 30\sin 25° - 9.8t$$
$$\mathbf{t = 1.294\ s}$$

Resolving horizontally (take right as +ve):

u = 30cos25°
s = r
a = 0
t = 2.588

Time to reach max height is 1.294, so total time until landing is 1.294 × 2 = 2.588

$$s = ut + \frac{1}{2}at^2$$
$$r = 30\cos 25° \times 2.588 + \frac{1}{2} \times 0 \times 2.588^2$$
$$= 70.4 \text{ m (to 3 s.f.)}$$

b) The first bit's easy: new h is just 8.2 + 1.5 = 9.7 m

Now use an equation of motion to work out the new horizontal distance:

s = –1.5
a = –9.8
u = 30sin25°
t = ?

$$s = ut + \frac{1}{2}at^2$$
$$-1.5 = (30\sin 25°)t - \frac{1}{2}(9.8)t^2$$
$$t^2 - 2.587t - 0.306 = 0$$
$$t = -0.11 \quad \text{or} \quad \mathbf{t = 2.70\ s}$$

Time can't be negative, so forget about this answer.

Resolving horizontally (take right as +ve)

s = r
u = 30cos25°
t = 2.70
a = 0

$$s = ut + \frac{1}{2}at^2$$
$$r = 30\cos 25° \times 2.70 + \frac{1}{2} \times 0 \times 2.7^2$$
$$= 73.4 \text{ m}$$

Practice Questions

1) ***A rifle fires a bullet horizontally at 120 ms^{-1}. The target is hit at a horizontal distance of 60 m from the end of the rifle. Find how far the target is vertically below the end of the rifle. Take g = 9.8 ms^{-2}.***

2) ***Sample exam question:***

> A stationary football is kicked with a speed of 20 ms^{-1}, at an angle of 30° to the horizontal, towards a goal 30 m away. The crossbar is 2.5 m above the level ground. Assuming the path of the ball is not impeded, determine whether the ball passes above or below the crossbar. Take g = 9.8 ms^{-2}. What assumptions does your model make? [6 marks]

Components make the world go round...

Using the equations of motion with projectiles is pretty much the same as before. The thing to remember is that horizontal acceleration is zero — great news because it makes the horizontal calculations as easy as a log-falling beginner's class.

Equations of Path

Skip these two pages if you're doing Edexcel or OCR A.

There are three simple equations to find the maximum height, time of flight and range on horizontal ground.

You've got to **Derive** the **Equations** — **Don't** just learn the results

Each time you use one of these path equations you have to show how the equation is derived.
If you just quote the final equation you'll get no marks.

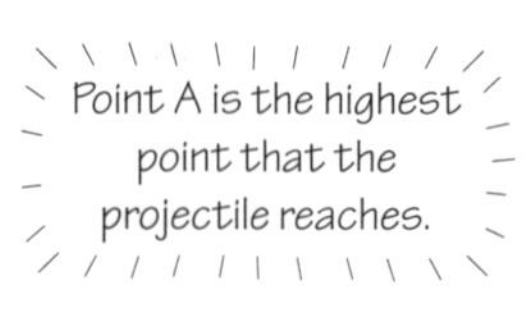

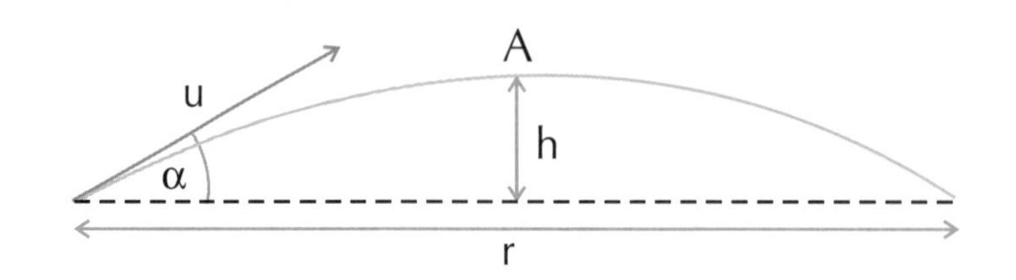

Maximum height, h

Resolving vertically:

s = h
a = –g
u = usinα
v = 0

$v^2 = u^2 + 2as$

$0 = (u\sin\alpha)^2 + 2(-g)h$

$$h = \frac{u^2\sin^2\alpha}{2g}$$

Time of Flight, T

Resolving vertically (to find time to A):

v = 0
u = usinα
t = ?
a = –g

$v = u + at$

$0 = u\sin\alpha^2 - gt$

$t = \dfrac{u\sin\alpha}{g}$

...but that's only to point A — you need to double t to get total time T:

$$T = \frac{2u\sin\alpha}{g}$$

Range, r

Resolving horizontally:

s = r
u = ucosα
$t = T = \dfrac{2u\sin\alpha}{g}$

$s = ut$ ← The horizontal acceleration is zero, so no ½at² term.

$r = \left(u\cos\alpha \times \dfrac{2u\sin\alpha}{g}\right)$

$= \dfrac{2u^2\sin\alpha\cos\alpha}{g}$

$$R = \frac{u^2\sin 2\alpha}{g}$$

Using sin2α = 2sinα cos α.

Example: A golf ball is struck at 30° to the horizontal with a speed of 40 ms⁻¹. Find the time of flight, the horizontal distance to where the ball first lands and the maximum height reached. Take g = 10 ms⁻².

Don't forget to derive the equations first (I'll miss out that step but you'd have to write out the derivations for the three equations above).

$$T = \frac{2u\sin\alpha}{g} = \frac{2\times40\times\sin 30}{10} = 4\text{ s}$$

$$r = \frac{u^2\sin 2\alpha}{g} = \frac{40^2\sin 60}{10} = 139\text{ m}$$

$$h = \frac{u^2\sin^2\alpha}{2g} = \frac{40^2\sin^2 30}{2\times10} = 20\text{ m}$$

If you ever get asked to work out when the range is largest, here's a handy tip:

r is at a maximum when α = 45°.

Equations of Path

Example: In the example on the last page, what is the nearest horizontal distance that a 10 m tall tree could be to the golfer if the ball is to go over it? What assumptions have you made in these examples? (Take $g = 10\ ms^{-2}$.)

Resolve vertically:

$u = 40\sin 30° = 20$
$a = -10$
$s = 10$
$t = ?$

$s = ut + \frac{1}{2}at^2$
$10 = 20t - 5t^2$
$5t^2 - 20t + 10 = 0$
$t = \frac{4 \pm \sqrt{8}}{2}$
$\mathbf{t = 0.586}$

Using the quadratic formula.

You get two values for t — the first is when the ball reaches 10 m from the ground on the way up, the second is when the ball reaches the same height on the way down. You need the first one.

Resolve horizontally:

$s = x$
$u = 40\cos 30°$
$t = 0.586$

$s = ut$
$x = 40\cos 30° \times 0.586$
$= 20.3\ m$

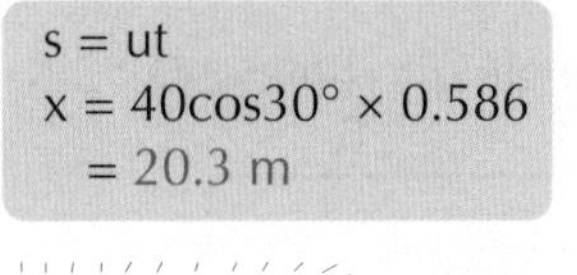

Assumptions:

- no air resistance
- no wind
- horizontal ground

Wind Speed only changes u_x

A following wind speed should be added to u_x, and you can leave u_y alone. Otherwise the method is the same as before.

Example: For the golfer above, how far does the ball travel before it lands if there is a tail wind of $3\ ms^{-1}$?

To find the time when the ball lands, put s = 0. So $u_y = 20$, $a = -10$, $s = 0$.

$s = u_y t + \frac{1}{2}at^2$
$0 = 20t - 5t^2$
$5t^2 = 20t$
$t = 0$ or $t = 4$.

t = 0 is when the ball is hit, so you need t = 4 here.

Now put t = 4 into your 'distance = speed × time' equation: $s = u_x t = (40\cos 30° + 3) \times 4 = 151\ m$

Practice Questions

1) A golf ball takes 4 seconds to land after being hit with a golf club. If it leaves the club with a speed of $22\ ms^{-1}$, at an angle of α to the horizontal, find α. Take $g = 9.8\ ms^{-2}$.

2) Sample exam question:

A cannon ball is fired at $50\ ms^{-1}$ at an angle of 25° to the horizontal. It hits its target 3 seconds later. Take $g = 9.8\ ms^{-2}$. Find:

a) The horizontal and vertical components of its initial velocity. [2 marks]
b) The position of the target relative to the cannon. [4 marks]
c) The maximum height reached by the ball. [2 marks]
d) The speed of the ball on impact with the target. [3 marks]

You'll never be a good golfer without path equations under your belt...

...and you'll have a hard time being an M1 mathematician too, come to think of it. OK, one more chorus and then let's call it a day... (to the tune of Auld Lang Syne) 'Derive the equations, don't just learn the results — then you'll get the marks.'

Projectile Trajectory Equations

Skip these two pages if you're doing Edexcel or OCR A.

Interestingly, the path of a projectile is a curve. You can find its equation in terms of its position (x,y). It's just a quadratic in x.

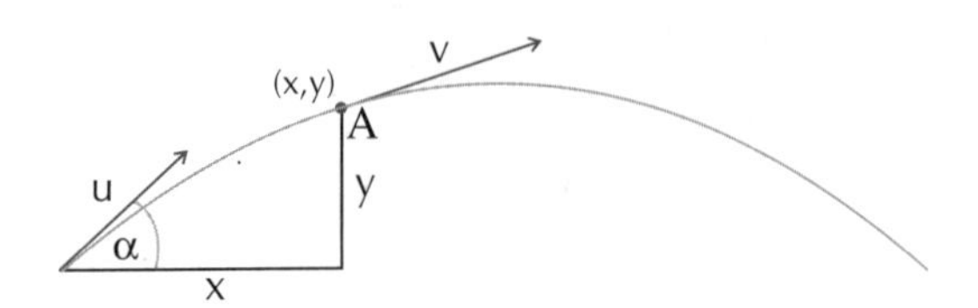

Learn the Derivations Below

Don't use these equations without showing where they come from — otherwise you won't get the marks.

For a projectile at A(x,y) given initial speed u at angle α to horizontal:

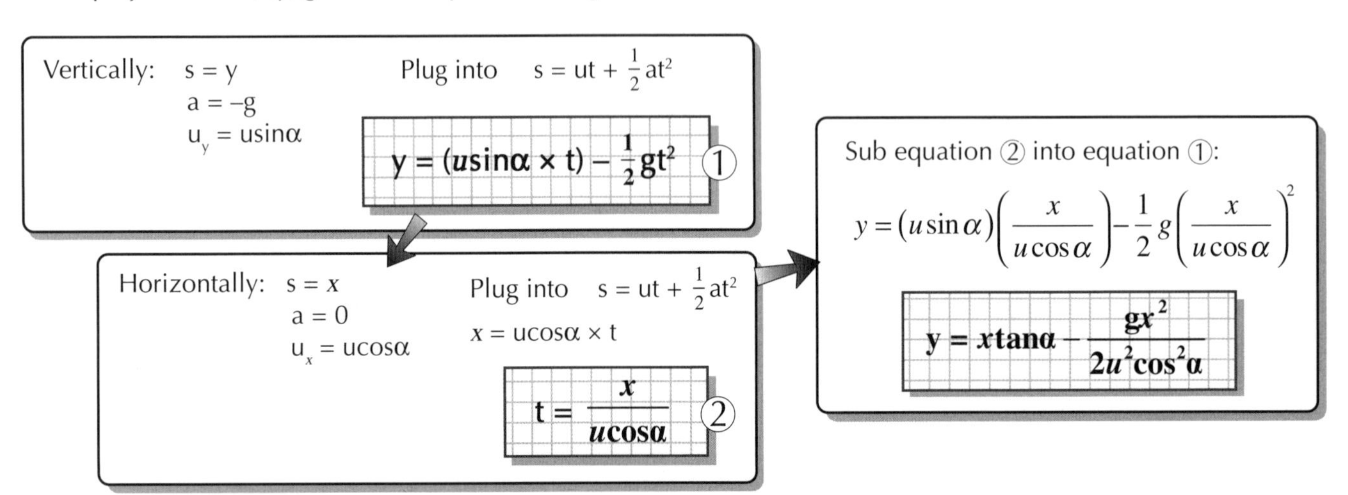

Doing this in the Exam should be a bit less scary as you will know some of the quantities as numbers.

You can also muck around with the third equation by using the identities $\left(\frac{1}{\cos^2\alpha}\right) = \sec^2\alpha$ and $\sec^2\alpha = (1 + \tan^2\alpha)$.

Example: A lawn sprinkler propels a droplet of water at a speed of 5 ms^{-1} at an angle α to the horizontal ground. Take $g = 10$ ms^{-2}.

a) Show that the vertical height, y, of a droplet at time t is given by $y = 5t(\sin\alpha - t)$

b) Find an expression for the horizontal distance, x, of a droplet at time t.

c) Find an equation in x and y for the path of a droplet.

d) Show that the range, r, can be related to the launch angle by $r^2\tan^2\alpha - 5r\tan\alpha + r^2 = 0$

e) Find α if the water reaches a distance of 2 m from the sprinkler.

a) [Diagram: projectile launched at speed 5 at angle α]

Resolving vertically: $u = 5\sin\alpha$, $s = y$, $a = -g = -10$, $t = ?$

$s = ut + \frac{1}{2}at^2$

$y = (5\sin\alpha \times t) - (\frac{1}{2} \times 10 \times t^2)$

$= 5t\sin\alpha - 5t^2$

$= \mathbf{5t(\sin\alpha - t)}$

b) Resolving horizontally: $u = 5\cos\alpha$, $s = x$, $t = ?$, $a = 0$

$s = ut + \frac{1}{2}at^2$

$x = 5\cos\alpha \times t$

$= \mathbf{5t\cos\alpha}$

Projectile Trajectory Equations

c) Rearrange the answer from b):

$$t = \left(\frac{x}{5\cos\alpha}\right)$$

Then sub into the answer from a):

$$y = 5t\sin\alpha - 5t^2$$

$$y = 5\left(\frac{x}{5\cos\alpha}\right)\sin\alpha - 5\left(\frac{x}{5\cos\alpha}\right)^2$$

$$\mathbf{y = x\tan\alpha - \frac{x^2}{5\cos^2\alpha}}$$

$\tan\alpha = \frac{\sin\alpha}{\cos\alpha}$

d) Use the identities $\frac{1}{\cos^2\alpha} = \sec^2\alpha = (1 + \tan^2\alpha)$ to rearrange the answer from c):

$$y = x\tan\alpha - \frac{1}{5}x^2(1 + \tan^2\alpha)$$

When droplet hits the ground y = 0, x = r:

$$0 = r\tan\alpha - \frac{1}{5}r^2 - \frac{1}{5}r^2\tan^2\alpha$$

$$0 = 5r\tan\alpha - r^2 - r^2\tan^2\alpha$$

$$\mathbf{r^2\tan^2\alpha - 5r\tan\alpha + r^2 = 0}$$

Multiplying everything by 5.

e) You need to find α when r = 2, so plug r = 2 into the formula from d):

$$4\tan^2\alpha - 10\tan\alpha + 4 = 0$$

$$2\tan^2\alpha - 5\tan\alpha + 2 = 0$$

$$(2\tan\alpha - 1)(\tan\alpha - 2) = 0$$

Dividing everything by 2 to make factorising easier.

Solve this quadratic in tanα:

$\tan\alpha = \frac{1}{2}$ or $\tan\alpha = 2$

$\alpha = 26.6^\circ$ or $\alpha = 63.4^\circ$

Practice Questions

1) A particle is projected with a speed of 20 ms^{-1}, at an angle α to the horizontal, from horizontal ground. It lands at a coordinate (x,y) relative to an origin at the point of projection (x and y measured in metres). Find α if the landing point is:

a) (30, 0) b) (40, 0) c) (50, 0) Take g = 10 ms^{-2}.

2) A particle is projected with a speed of 20 ms^{-1} at an angle α to the horizontal. Find the possible values of α (to 1 d.p.) if the particle passes through the point with coordinates (12, 6) relative to an origin at the point of projection. Distances are in metres. Take g = 10 ms^{-2}.

3) Sample exam question:

A projectile is fired at an angle of 40° to the horizontal with speed u ms^{-1}. It passes through the point A with coordinate (50, 30), in metres, relative to an origin at the point of projection. Find the magnitude of u and the time elapsed when the projectile reaches A. Take g = 10 ms^{-2}. [6 marks]

And now for an entirely unhelpful tip...

To round off the last section, here's my list of top twelve films about mattresses:
1) The Invisible Mattress 2) The Mattress of King George 3) The Mattress Strikes Back 4) Dial 'M' for Mattress
5) Gone with the Mattress 6) Lock, Stock and Two Smoking Mattresses 7) Mattress of the Apes
8) There's Something About Mattress 9) Silence of the Mattresses 10) The Mattress Reloaded
11) Star Wars Episode 1: The Phantom Mattress 12) Indiana Jones and the Mattress of Doom

Answers

Section One — Kinematics

Page 3

1) $u = 3;\ v = 9;\ a = a;\ s = s;\ t = 2$
Use $s = ½(u + v)t$
$s = ½(3 + 9) \times 2$
$s = ½(12) \times 2 = 12\ m$

2) $u = 3;\ v = 0;\ a = -9.8;\ s = s;\ t = t$
[$v = 0$ because when projected objects reach the top of their motion, they STOP momentarily, then come down again.]
[$a = -9.8\ ms^{-2}$ because gravity will SLOW the ball down, so it's negative.]
Use $v = u + at$
$0 = 3 + (-9.8)t$
$0 = 3 - 9.8t$
$9.8t = 3$
$t = \dfrac{3}{9.8} = 0.31\ s$

3)a) $u = u;\ v = v;\ a = 10;\ s = 4h;\ t = 1.2$
You need an equation with u and s because they're in the equation you have to show is true. You know t and a, so the equation to use is one with u, a, s and t in it – that's $s = ut + ½ at^2$.
$4h = 1.2u + ½ \times 10 \times 1.2^2$ [1 mark]
$4h = 1.2u + 7.2$ [1 mark]

b) i) You have to think of the entire fall through 8 floors, because the question demands we have u involved.
$u = u;\ v = v_1;\ a = 10;\ s = 8h;\ t = 1.8$
[$s = 8h$ because it's 4 floors + 4 floors]
[$t = 1.8$ because it's $1.2 + 0.6$]
Use $s = ut + ½ at^2$ [1 mark]
$8h = 1.8u + ½ \times 10 \times 1.8^2$
$8h = 1.8u + 16.2$ [1 mark]

ii) You now have two equations, both with h and u in, so solve simultaneously.
The easiest way to do this is to double the equation from a):
$8h = 2.4u + 14.4$
and subtract the equation from b)i) from this:
$0 = 0.6u - 1.8$
$0.6u = 1.8$
$u = 3\ ms^{-1}$ [1 mark]
Substitute this value back into the equation from a):
$4h = 1.2 \times 3 + 7.2$
$4h = 10.8$
$h = 2.7\ m$ [1 mark]

c) Modelling assumption made is that the sandwich is a particle, so there's no air resistance on it. [1 mark]

Page 5

1)

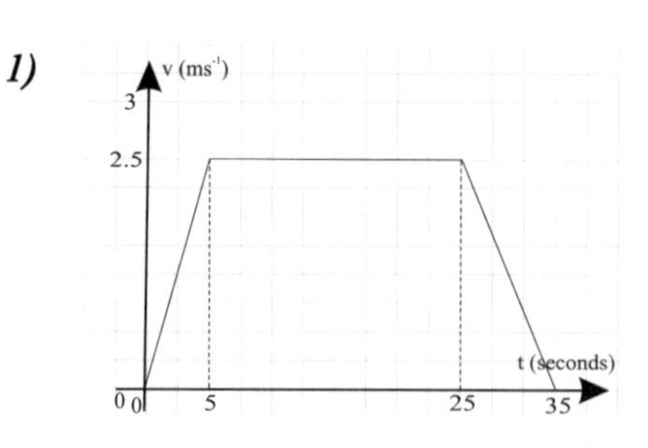

distance $= (5 \times 2.5) \div 2 + (20 \times 2.5) + (10 \times 2.5) \div 2$
$= 68.75\ m$

2) velocity = area under (t,a) graph
a) $t = 3$, area $= (3 \times 5) \div 2 = 7.5\ ms^{-1}$
b) $t = 5$, area $= 7.5 + (2 \times 5) = 17.5\ ms^{-1}$
c) $t = 6$, area $= 17.5 + (1 \times 5) \div 2 = 20\ ms^{-1}$

3)a) Between $t = 10$ and $t = 15$. The graph is horizontal at this point. [1 mark]
b) $50 \div 10 = 5\ ms^{-1}$ [2 marks]
c) $50\ m \div 20\ s = 2.5\ ms^{-1}$ back towards the starting point. [2 marks]

Page 7

1) a) Multiplying out gives $v = 15t^2 - 2t^3$
Differentiating v with respect to t gives $a = 30t - 6t^2$ [1 mark]
If $a = 0$:
$30t - 6t^2 = 0$
$6t(5 - t) = 0$
So $t = 0$ or $t = 5$ [2 marks]

b) Integrating v with respect to t between $t = 0$ and $t = 4$ gives r, the distance the particle has moved.

$$r = \int_0^4 \left(15t^2 - 2t^3\right)dt = \left[5t^3 - \frac{1}{2}t^4\right]_0^4 \quad \text{[2 marks]}$$

so $r = 320 - 128 = 192$ [2 marks]

2)a) Integrating v gives the position vector $\mathbf{r}$:

$$\mathbf{r} = \begin{bmatrix} 3t^2 - 2t + c_1 \\ 5t + c_2 \end{bmatrix} \quad \text{[2 marks]}$$

When $t = 0$, $\mathbf{r} = \begin{bmatrix} 4 \\ 3 \end{bmatrix}$ [1 mark]

so $c_1 = 4$ and $c_2 = 3$, giving $\mathbf{r} = \begin{bmatrix} 3t^2 - 2t + 4 \\ 5t + 3 \end{bmatrix}$ [2 marks]

b) Differentiating v with respect to t gives $\mathbf{a} = \begin{bmatrix} 6 \\ 0 \end{bmatrix}$ [2 marks]

Answers

Section Two — Vectors

Page 9

1) *Displacement* = $(15 \times 0.25) - (10 \times 0.75) = -3.75$ km
Time taken = 1 hour
Average velocity = -3.75 km/h (i.e. 3.75 kmh^{-1} south)

2) $(3\mathbf{i} + 7\mathbf{j}) + 2 \times (-2\mathbf{i} + 2\mathbf{j}) - 3 \times (\mathbf{i} - 3\mathbf{j})$
$= (3 - 4 - 3)\mathbf{i} + (7 + 4 + 9)\mathbf{j}$
$= -4\mathbf{i} + 20\mathbf{j}$

3) *Horizontally:* $0 + 5\cos 30 = 4.33$ N [1 mark]
Vertically: $4 - 5\sin 30 = 1.5$ N [1 mark]

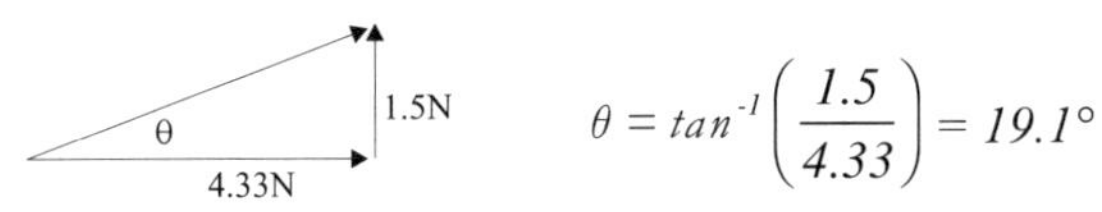

$\theta = \tan^{-1}\left(\frac{1.5}{4.33}\right) = 19.1°$

i.e. $\theta = 19.1°$ above the horizontal [1 mark]

Magnitude = $\sqrt{1.5^2 + 4.33^2} = 4.58$ N [2 marks]

Page 11

1)

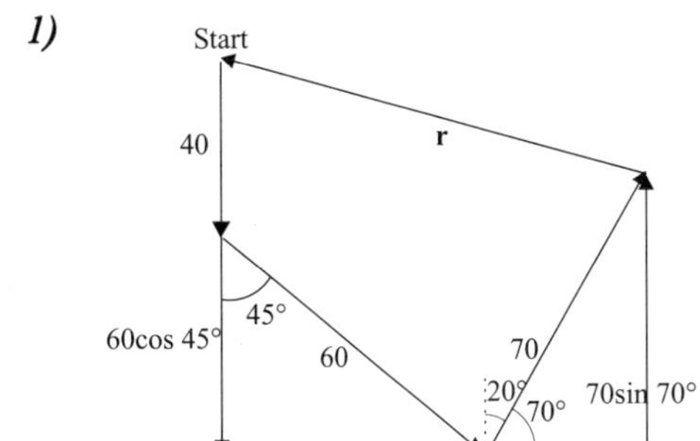

Resolving East: $60\sin 45° + 70\cos 70° = 66.4$ miles
Resolving North: $-40 - 60\cos 45° + 70\sin 70°$
$= -16.6$ miles

Magnitude of $\mathbf{r}$ = $\sqrt{66.4^2 + 16.6^2} = 68.4$ miles

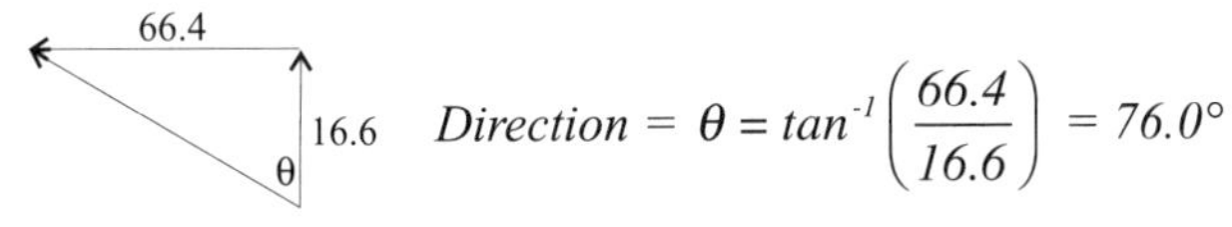

Direction = $\theta = \tan^{-1}\left(\frac{66.4}{16.6}\right) = 76.0°$

Bearing is $360° - 71.8° = 284°$

2)

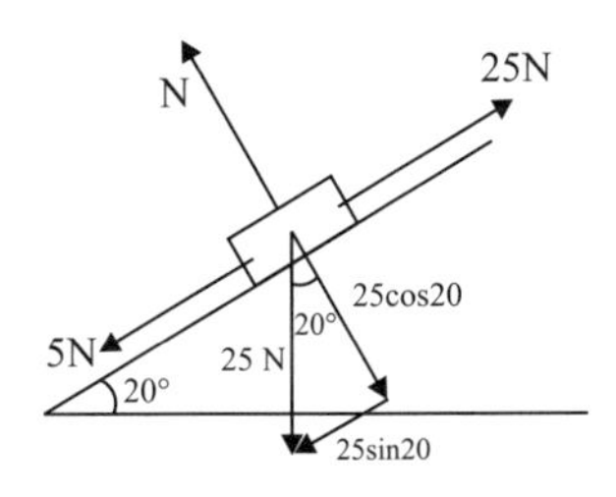

Force perpendicular to the slope: $N = 25\cos 20° = 23.5$ N
Force parallel to the slope: $25 - 25\sin 20° - 5 = 11.4$ N.
So the resultant force is 11.4 N up the slope.

3) [2 marks for diagram]

Magnitude = $\sqrt{2^2 + 3^2} = \sqrt{13} = 3.61$ ms^{-1} [2 marks]

$\theta = \tan^{-1}\left(\frac{3}{2}\right) = 56.3°$ [1 mark]

So angle to river bank is $90° - 56.3° = 33.7°$ [1 mark]

Section Three — Statics

Page 13

1)a) Small point mass, no air resistance, no wind, released from rest

b) Small point mass, no air resistance, no wind, released from rest
The drinks can is very smooth so ignoring air resistance is OK.

c) Same assumptions as in a) and b), (although realistically it might not be safe to ignore wind if you're outside).

2) Assumptions: Point mass, one point of contact to ground, constant driving force D from engine, constant friction, F, includes road resistance and air resistance, acceleration = 0 as it's moving at 25mph.

3) Assumptions: Point mass, can ignore friction (air) upwards as parachute isn't open yet.

Page 15

1) a) $R = \sqrt{4^2 + 3^2} = 5$ N

$\tan\theta = \frac{4}{3}$

$\theta = 53.1°$

b) $R = \sqrt{(8 + 5\cos 60)^2 + (5\sin 60)^2} = 11.4$ N

$\tan\theta = \frac{5\sin 60}{8 + 5\cos 60} = 0.412$ so $\theta = 22.4°$

c) *Total force up* $= 6 - 4\sin 10° - 10\sin 20°$
$= 1.885$ N
Total force left $= 10\cos 20° - 4\cos 10°$
$= 5.458$ N

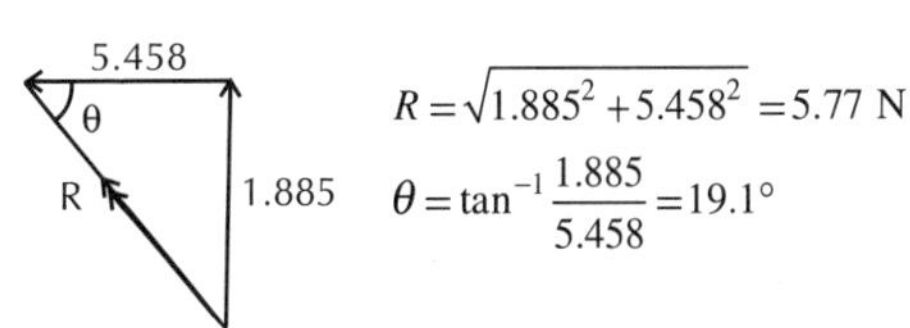

$R = \sqrt{1.885^2 + 5.458^2} = 5.77$ N

$\theta = \tan^{-1}\frac{1.885}{5.458} = 19.1°$

Answers

2)a)

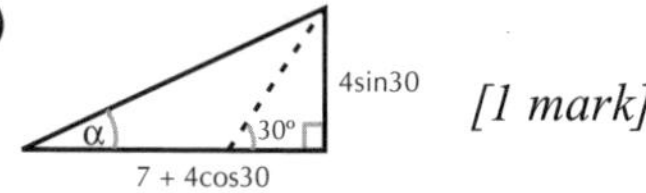

[1 mark]

$R = \sqrt{(4\sin 30°)^2 + (7 + 4\cos 30°)^2}$ *[1 mark]*

$= 10.7$ N *[1 mark]*

b) $\tan\alpha = \dfrac{4\sin 30°}{7 + 4\cos 30°} = 0.1911$ *[1 mark for tanα, 1 mark for the rest]*

$\alpha = 10.8°$ *[1 mark]*

3)

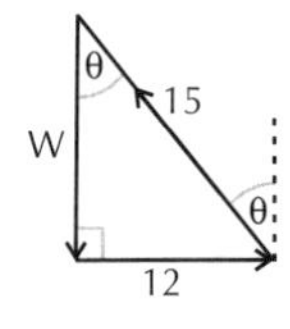

a) $\sin\theta = \dfrac{12}{15}$ *[1 mark]*

so $\theta = 53.1°$ *[1 mark]*

b) $15^2 = W^2 + 12^2$ *[1 mark]*

$W = \sqrt{15^2 - 12^2} = 9$ N *[1 mark]*

Remove W and the particle moves in the opposite direction to W, i.e. upwards. This resultant of the two remaining forces is 9 N [1 mark] upwards [1 mark].

Page 17

1)

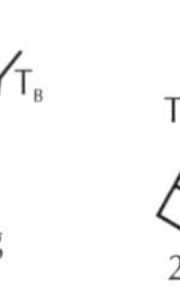

a) $\tan 30 = \dfrac{20}{T_B}$

$T_B = \dfrac{20}{\tan 30} = 34.6$ N

b) $\sin 30 = \dfrac{20}{Mg}$

$Mg = \dfrac{20}{\sin 30}$

$M = 4.08$ kg

2)

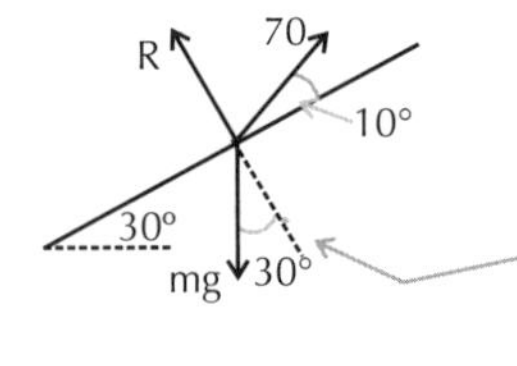

Huge Hint: The angle of the plane to the horizontal (30°) will always be the angle in here

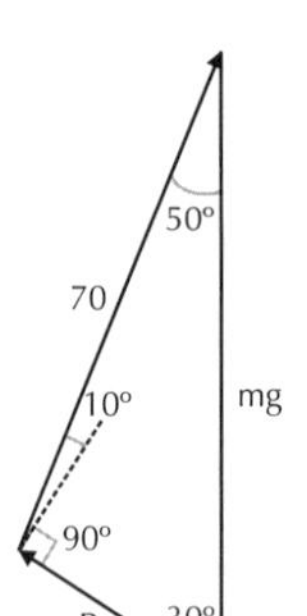

Sine rule: $\dfrac{mg}{\sin 100} = \dfrac{70}{\sin 30}$

So $mg = \dfrac{70\sin 100}{\sin 30}$

$m = 14.1$ kg

$\dfrac{R}{\sin 50} = \dfrac{70}{\sin 30}$

So $R = \dfrac{70\sin 50}{\sin 30} = 107$ N

3)

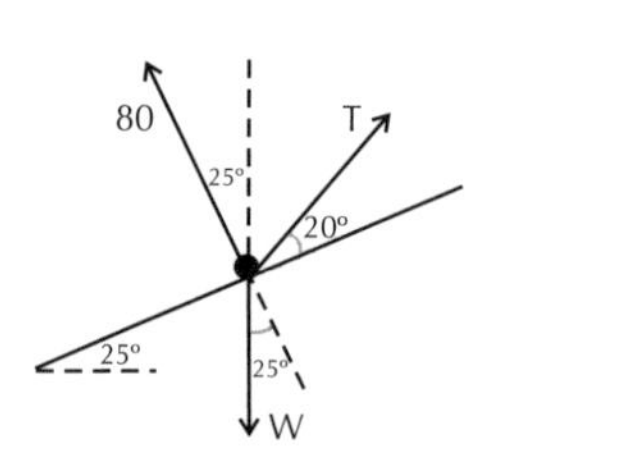

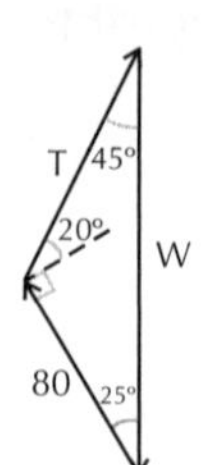

[2 marks for method and diagram]

a) $\dfrac{T}{\sin 25°} = \dfrac{80}{\sin 45°}$ *[1 mark]*

So $T = \dfrac{80\sin 25°}{\sin 45°} = 47.8$ N *[1 mark]*

b) $\dfrac{W}{\sin 110°} = \dfrac{80}{\sin 45°}$ *[1 mark]*

So $W = \dfrac{80\sin 110°}{\sin 45°} = 106$ N *[1 mark]*

Page 19

1) Resolve vertically: $R = 12g$

Use formula: $F \le \mu R$

$F \le \frac{1}{2}(12g)$

$F \le 58.8$ N

50 N isn't big enough to overcome friction — so it doesn't move.

2) *Force would have to be > 58.8 N*

3) $\tan\alpha = \mu = 0.2$ *[1 mark]*

So $\alpha = \tan^{-1} 0.2 = 11.3°$ *[1 mark]*

4) *Resolve horizontally*: $S\cos 40° = F$ *[1 mark]*

Resolve vertically: $R = 2g + S\sin 40°$ *[1 mark]*

It's limiting friction so $F = \mu R$ *[1 mark]*

So $S\cos 40° = \dfrac{3}{10}(2g + S\sin 40°)$

$S\cos 40° = 0.6g + 0.3S\sin 40°$

$S\cos 40° - 0.3S\sin 40° = 0.6g$

$S(\cos 40° - 0.3\sin 40°) = 0.6g$ *[1 mark]*

$S = 10.3$ N *[1 mark]*

Answers

Page 21

1) *Moments about B:* $60g \times 3 = T_2 \times 8$

$$\text{So } T_2 = \frac{180g}{8} = 220.5 \text{ N}$$

Vertically balanced forces, so $T_1 + T_2 = 60g$

$$T_1 = 367.5 \text{ N}$$

2)

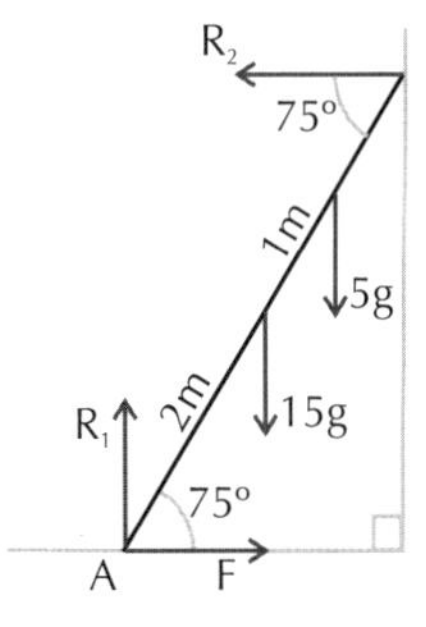

[2 marks]

a) *Moments about A:*

$15g \times 2\cos 75° + 5g \times 3\cos 75° = R_2 \times 4\sin 75°$ *[1 mark]*

So $R_2 = \dfrac{45g\cos 75°}{4\sin 75°}$ *[1 mark]*

$= 29.5$ N *[1 mark]*

b) *Resolving vertically:* $R_1 = 15g + 5g = 196$ N *[2 marks]*

c) *Resolving horizontally:* $F = R_2 = 29.5$ N *[1 mark]*

$F \leq \mu R_1$

$29.5 \leq \mu \times 196$ *[1 mark]*

$\mu \geq \dfrac{29.5}{196}$

So $\mu \geq 0.15$ *[1 mark]*

Page 23

1)

$M\bar{x} = \sum m_i x_i$

$12\bar{x} = (2\times 0) + (4\times 0) + (6\times 3)$

$\bar{x} = 1.5$ m

$M\bar{y} = \sum m_i y_i$

$12\bar{y} = (2\times 0) + (4\times 2) + (6\times 2)$

$\bar{y} = 1\frac{2}{3}$ m

2)

$Area_1 = 5\times 1 = 5m^2$

$y_1 = 2m$

$Area_2 = 2\times 1.5 = 3m^2$

$y_2 = 0.75m$

$M\bar{y} = \sum m_i y_i$

$(5+3)\bar{y} = 5\times 2 + 3\times 0.75$

$\bar{y} = 1.53m$ above A

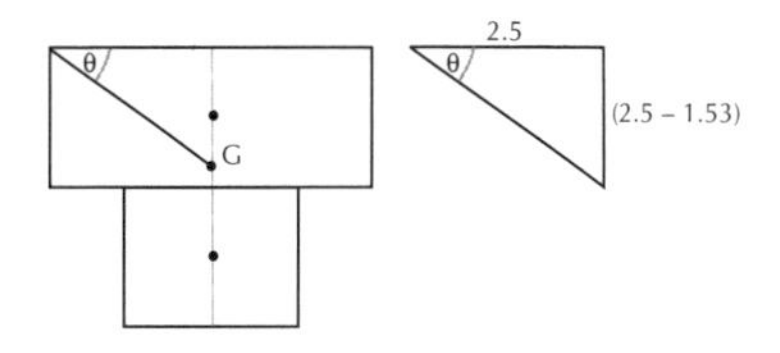

$\tan\theta = \dfrac{0.97}{2.5}$

So $\theta = 21.2°$

So angle to horizontal

is $90 - 21.2 = 68.8°$

3)

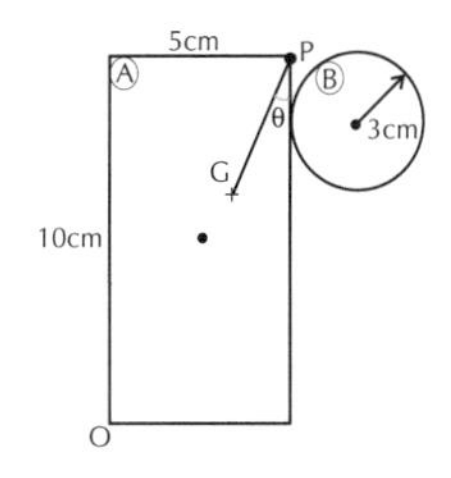

$M_1 = \text{Area A} = 5\times 10 = 50 \text{ cm}^2$

$x_1 = 2.5cm \quad y_1 = 5cm$

$M_2 = \text{Area B} = \pi \times 3^2 = 9\pi \text{ cm}^2$

$x_2 = 8cm \quad y_2 = 7cm$

$M\bar{x} = \sum m_i x_i$

$(50 + 9\pi)\bar{x} = 50\times 2.5 + 9\pi \times 8$

So $\bar{x} = 4.487\,cm$ *[2 marks]*

$M\bar{y} = \sum m_i y_i$

$(50 + 9\pi)\bar{y} = 50\times 5 + 9\pi \times 7$

So $\bar{y} = 5.722\,cm$ *[2 marks]*

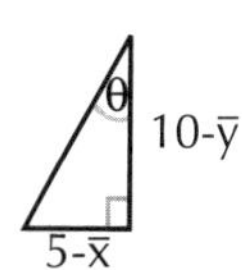

$\tan\theta = \dfrac{0.513}{4.278}$

So $\theta = 6.8°$ *[2 marks]*

Answers

Section Four — Dynamics

Page 25

1)

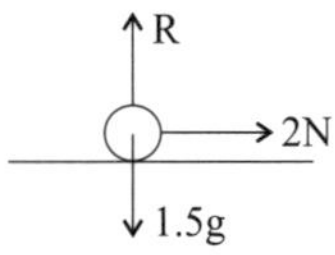

Resolve horizontally: $F_{net} = ma$

$$2 = 1.5a \quad so \quad a = 1\tfrac{1}{3}\ ms^{-2}$$

$$v = u + at$$

$$v = 0 + (1\tfrac{1}{3} \times 3) = 4\ ms^{-1}$$

2) $(24\mathbf{i} + 18\mathbf{j}) = \begin{pmatrix} 24 \\ 18 \end{pmatrix}$

$(6\mathbf{i} + 22\mathbf{j}) = \begin{pmatrix} 6 \\ 22 \end{pmatrix}$

$$F_{net} = \begin{pmatrix} 24 \\ 18 \end{pmatrix} + \begin{pmatrix} 6 \\ 22 \end{pmatrix} = \begin{pmatrix} 30 \\ 40 \end{pmatrix} = 30\mathbf{i} + 40\mathbf{j}$$

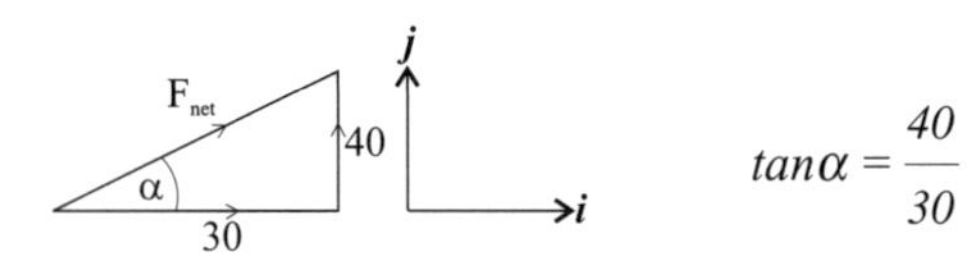

$\tan\alpha = \dfrac{40}{30}$

So $\alpha = 53.1°$

$F_{net} = \sqrt{30^2 + 40^2} = 50\ N$

Resolve in direction of F_{net}:

$F_{net} = ma$

$50 = 8a$ so $a = 6.25\ ms^{-2}$

$s = ut + \frac{1}{2}at^2$

$s = 0 \times 3 + \frac{1}{2} \times 6.25 \times 3^2 = 28.1\ m$

3)

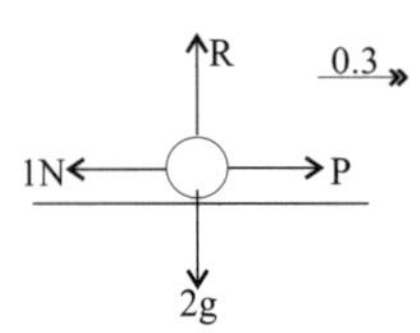

Resolve horizontally: $F_{net} = ma$

$P - 1 = 2 \times 0.3$

$P = 1.6\ N$

Resolve vertically: $R = 2g$

Limiting friction: $F = \mu R$

$1 = \mu \times 2g$

So $\mu = 0.05$ *(to 2 d.p.)*

4)a)

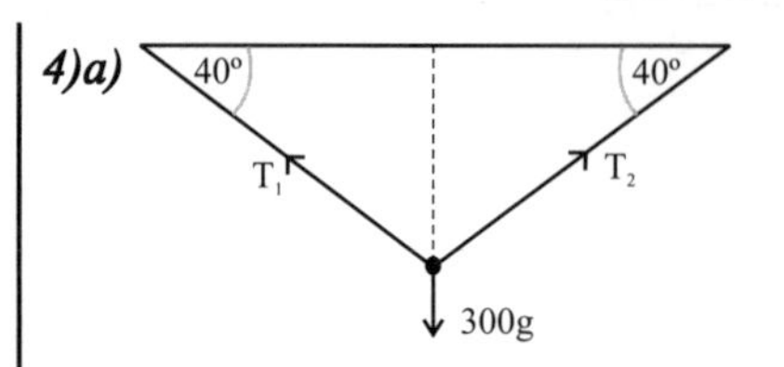

Constant velocity, so $a = 0$

Resolve horizontally: $F_{net} = ma$

$T_2\cos40 - T_1\cos40 = 300 \times 0$

$T_2\cos40 = T_1\cos40$

$T_2 = T_1$ *(1 mark)*

Resolve vertically: $F_{net} = ma$

$T_1\sin40 + T_2\sin40 - 300g = 300 \times 0$ *(1 mark)*

Let $T_1 = T_2 = T$: $2T\sin40 = 300g$ *(1 mark)*

$T = 2290\ N$ *(to 3 s.f.) (1 mark)*

b) Resolve horizontally: $F_{net} = ma$

$T_2\cos40 - T_1\cos40 = 300 \times 0.4$ *(1 mark)*

$T_2 - T_1 = 156.65\ N$ ① *(1 mark)*

Resolve vertically: $F_{net} = ma$

$T_1\sin40 + T_2\sin40 - 300g = 300 \times 0$ *(1 mark)*

$T_1\sin40 + T_2\sin40 = 300g$

So $T_1 + T_2 = 4573.83\ N$ ② *(1 mark)*

from ①: $T_2 = T_1 + 156.65$

into ②: $T_1 + (T_1 + 156.65) = 4573.83$

so $2T_1 = 4417.17$

$T_1 = 2210\ N$ *(to 3 s.f.) (1 mark)*

So $T_2 = 2370\ N$ *(to 3 s.f.) (1 mark)*

c) Modelling assumptions: particle is considered as a point mass, there's no air resistance, it's a constant acceleration, etc. (1 mark each for any 2)

Page 27

1)

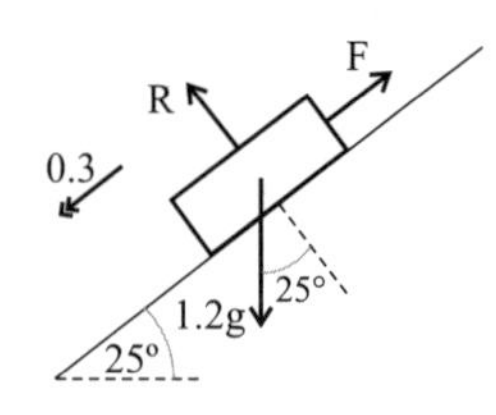

Resolving in ↖ *direction:* $F_{net} = ma$

$R - 1.2g\cos25° = 1.2 \times 0$

$R = 1.2g\cos25°$

$R = 10.66\ N$

Resolving in ↙ *direction:* $F_{net} = ma$

$1.2g\sin25° - F = 1.2 \times 0.3$

So $F = 1.2g\sin25 - 1.2 \times 0.3$

$= 4.61\ N$

Limiting friction, so: $F = \mu R$

$4.61 = \mu \times 10.66$

$\mu = 0.43$ *(to 2 d.p.)*

Answers

Assumptions: any two from: i) brick slides down line of greatest slope; ii) acceleration is constant; iii) no air resistance; iv) point mass / particle

2)

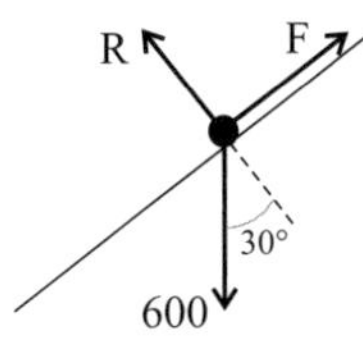

$\mu = 0.5$

Resolving in ↖ *direction:* $F_{net} = ma$

$$R - 600\cos 30° = \left(\frac{600}{g}\right) \times 0$$

$$R = 600\cos 30°$$

Sliding, so $F = \mu R$

$$F = 0.5 \times 600\cos 30° = 259.8$$

Resolving in ↙ *direction:* $600\sin 30° - F = \left(\frac{600}{g}\right)a$

$$600\sin 30 - 259.8 = 61.22 \times a$$

$$a = 0.656\ ms^{-2}$$

$$\left.\begin{array}{r} u = 0 \\ s = 20 \\ a = 0.656 \\ v = ? \end{array}\right\} \quad \begin{array}{l} v^2 = u^2 + 2as \\ v^2 = 0^2 + 2\times 0.656\times 20 \\ v^2 = 26.26 \\ v = 5.12\ ms^{-1} \end{array}$$

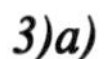

3)a)

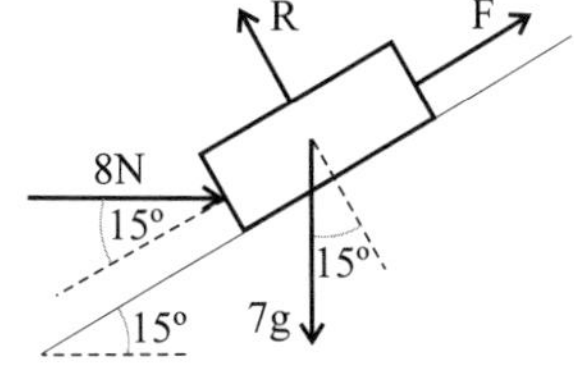

$F_{net} = ma$

Resolving in ↗ *direction:* $8\cos 15 + F - 7g\sin 15 = 7 \times 0$

$$F = 7g\sin 15 - 8\cos 15$$

$$= 10.03\ N \quad \textit{[2 marks]}$$

Resolving in ↖ *direction:* $F_{net} = ma$

$$R - 8\sin 15 - 7g\cos 15 = 7 \times 0 \quad \textit{[1 mark]}$$

$$R = 8\sin 15 + 7g\cos 15$$

$$= 68.33\ N \quad \textit{[1 mark]}$$

Limiting friction: $F = \mu R$

$$10.03 = \mu \times 68.33$$

$$\mu = 0.15 \quad \textit{[2 marks]}$$

b) *8 N removed:*

Resolving in ↙ *direction:* $7g\sin 15 - F = 7a$ ① *[1 mark]*

Resolving in ↖ *direction:* $R - 7g\cos 15 = 7 \times 0$

$$R = 7g\cos 15$$

$$= 66.26\ N \quad \textit{[1 mark]}$$

$$F = \mu R$$

$$F = 0.15 \times 66.26 = 9.939\ N \quad \textit{[2 marks]}$$

①: $7g\sin 15° - 9.939 = 7a$

$a = \frac{7.82}{7} = 1.12\ ms^{-2}$ *(to 2d.p.) [1 mark]*

$s = 3;\ u = 0;\ a = 1.12;\ t = ?$

$s = ut + \frac{1}{2}at^2$

$3 = 0 + \frac{1}{2} \times 1.12 \times t^2$

$t = \sqrt{\frac{6}{1.12}} = 2.3\ s$ *[2 marks]*

Page 29

1)a) $\mathbf{u} = \begin{pmatrix} 0 \\ -6 \end{pmatrix};\quad \mathbf{v} = \begin{pmatrix} 8 \\ 0 \end{pmatrix};\quad t = 20;\quad \mathbf{a} = ?$

$\mathbf{v} = \mathbf{u} + \mathbf{a}t$

$$\begin{pmatrix} 8 \\ 0 \end{pmatrix} = \begin{pmatrix} 0 \\ -6 \end{pmatrix} + 20\mathbf{a}$$

$$\mathbf{a} = \begin{pmatrix} 0.4 \\ 0.3 \end{pmatrix} = 0.4\mathbf{i} + 0.3\mathbf{j}$$

$$\mathbf{a} = \sqrt{0.3^2 + 0.4^2} = 0.5\ ms^{-2}$$

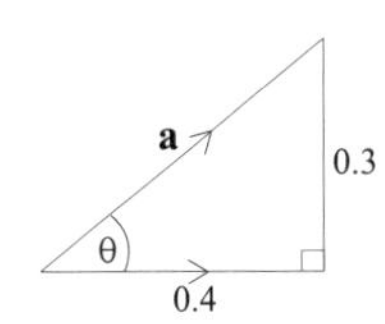

$$\theta = \tan^{-1}\left(\frac{0.3}{0.4}\right) = 36.9°$$

Bearing $= 90° - 36.9° = 053.1°$

So $\mathbf{a} = 0.5\ ms^{-2}$ *on bearing* $053.1°$

b) $\mathbf{s} = \mathbf{u}t + \frac{1}{2}\mathbf{a}t^2$

$$\begin{pmatrix} x \\ y \end{pmatrix} = \begin{pmatrix} 0 \\ -6 \end{pmatrix}5 + \frac{1}{2}\begin{pmatrix} 0.4 \\ 0.3 \end{pmatrix}5^2 = \begin{pmatrix} 5 \\ -26 \end{pmatrix} \quad \textit{(to 2 s.f.)}$$

At $t = 5$, $P(5, -26)$

2)a) $\mathbf{r} = \begin{pmatrix} 15t + 10 \\ 15\sqrt{3}t - 5t^2 \end{pmatrix} \quad t > 1$

Plug in $t = 3$: $\mathbf{r} = \begin{pmatrix} 55 \\ 32.9 \end{pmatrix}$ *[1 mark]*

So $d = \sqrt{32.9^2 + 55^2} = 64\ m$ *(to 2 s.f.) [1 mark]*

b)

t	1	2	3
x	25	40	55
y	21	32	33

(values to 2 s.f.) [2 marks]

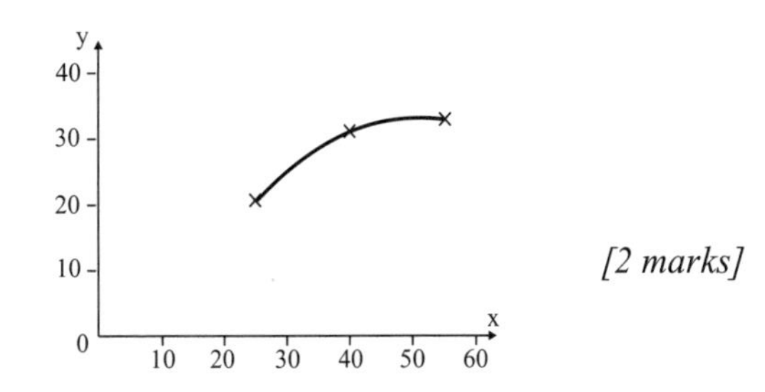

[2 marks]

Answers

c) $\begin{pmatrix} x \\ y \end{pmatrix} = \begin{pmatrix} 15t+10 \\ 15\sqrt{3}t - 5t^2 \end{pmatrix}$ so $x = 15t + 10$

So $t = \left(\frac{x-10}{15}\right)$ *[1 mark]*

Plug into y: $y = 15\sqrt{3}\left(\frac{x-10}{15}\right) - 5\left(\frac{x-10}{15}\right)^2$ *[1 mark]*

$y = \sqrt{3}x - 10\sqrt{3} - \frac{1}{45}(x^2 - 20x + 100)$

$y = -\frac{1}{45}x^2 + (\sqrt{3} + \frac{4}{9})x - (\frac{20}{9} + 10\sqrt{3})$ *[1 mark]*

$y = 2.2x - 20 - 0.02x^2$ *[1 mark]*

d) $\mathbf{r} = \begin{pmatrix} 15t+10 \\ 15\sqrt{3}t - 5t^2 \end{pmatrix}$ $\mathbf{v} = \frac{d\mathbf{r}}{dt} = \begin{pmatrix} 15 \\ 15\sqrt{3} - 10t \end{pmatrix}$ *[2 marks]*

$\mathbf{a} = \frac{d\mathbf{v}}{dt} = \begin{pmatrix} 0 \\ -10 \end{pmatrix}$ *[1 mark]*

e) For large t, $y < 0$ — i.e. the aircraft crashes!

Page 31

1) Taking tractor and trailer together (and calling the resistance force on the trailer R):

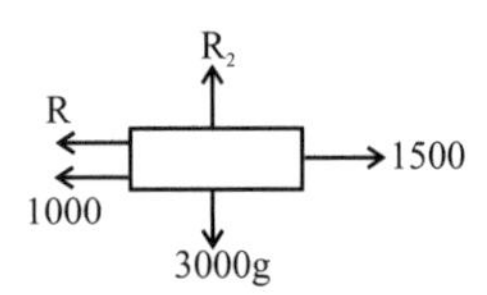

Resolving horizontally: $F_{net} = ma$

$1500 - R - 1000 = 3000 \times 0$

$R = 500\ N$

For trailer alone:

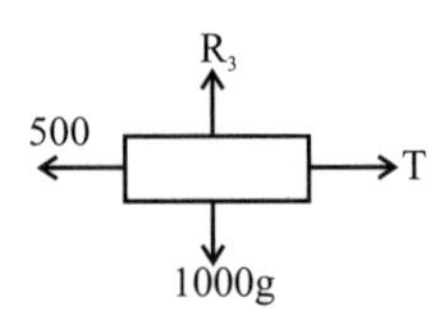

Resolving horizontally: $F_{net} = ma$

$T - 500 = 1000 \times 0$

$T = 500\ N$

T could be found instead by looking at tractor forces horizontally.

2)

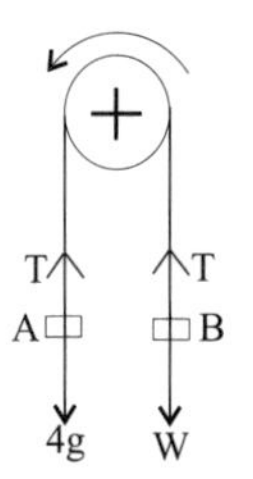

Resolving downwards for A: $F_{net} = ma$

$4g - T = 4 \times 1.2$

$T = 4g - 4.8$ ①

Resolving upwards for B: $F_{net} = ma$

$T - W = \frac{W}{g} \times 1.2$ ②

Sub ① into ② : $(4g - 4.8) - W = \frac{W}{g}(1.2)$

$4g - 4.8 = W(1 + \frac{1.2}{g})$

So $W = 30.6\ N$

3 a)

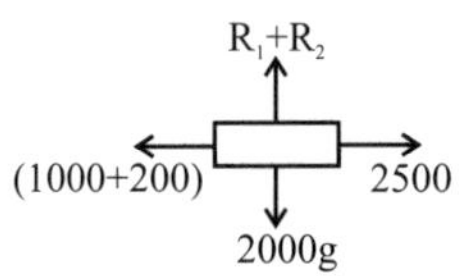

Resolving horizontally: $F_{net} = ma$

$2500 - 1200 = 2000a$

$a = 0.65\ ms^{-2}$

b) Either:

Caravan

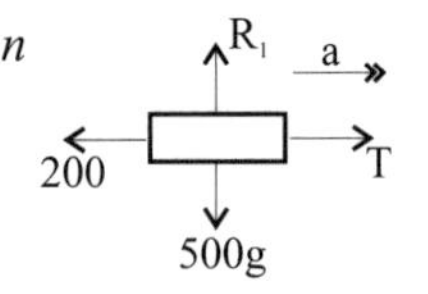

Resolving horizontally: $F_{net} = ma$

$T - 200 = 500 \times 0.65$ *[2 marks]*

$T = 525\ N$ *[1 mark]*

Or:

Car

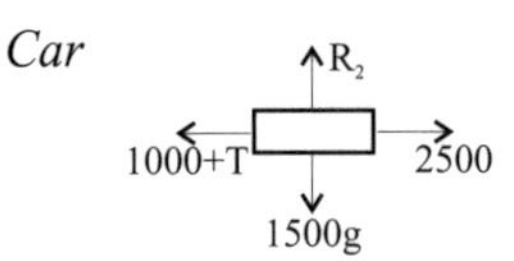

Resolving horizontally: $F_{net} = ma$

$2500 - (1000 + T) = 1500 \times 0.65$

$2500 - 1000 - T = 975$ *[2 marks]*

$1500 - 975 = T$

$T = 525\ N$ *[1 mark]*

Answers

Page 33

1)

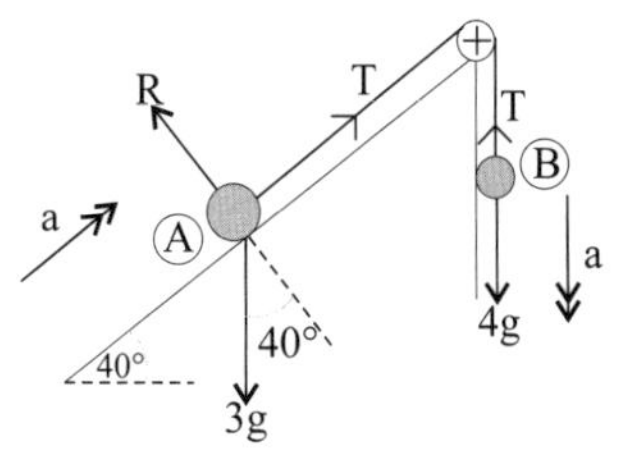

For B, resolving vertically: $F_{net} = ma$

$4g - T = 4a$

$T = 4g - 4a$ ①

For A, resolving in ↗ *direction:* $F_{net} = ma$

$T - 3g\sin 40 = 3a$ ②

Sub ① into ② : $4g - 4a - 3g\sin 40 = 3a$

$4g - 3g\sin 40 = 7a$

$a = 2.9\ ms^{-2}$

Sub into ① : $T = 4g - (4 \times 2.9) = 27.6\ N$ *(to 3 s.f.)*

If equilibrium, then for B: $T = 4g$

Then for Ⓐ

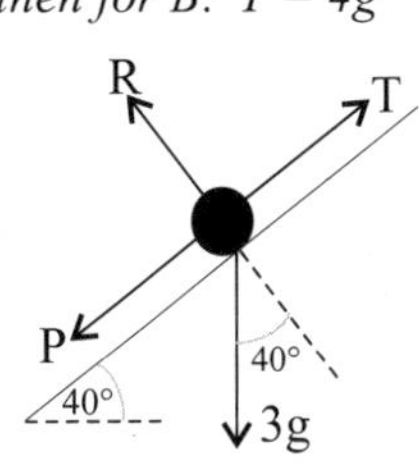

Resolving in ↗ *direction:* $F_{net} = ma$

$T - 3g\sin 40 - P = 0$

$4g - 3g\sin 40 = P$

$P = 20.3N$

2)a)

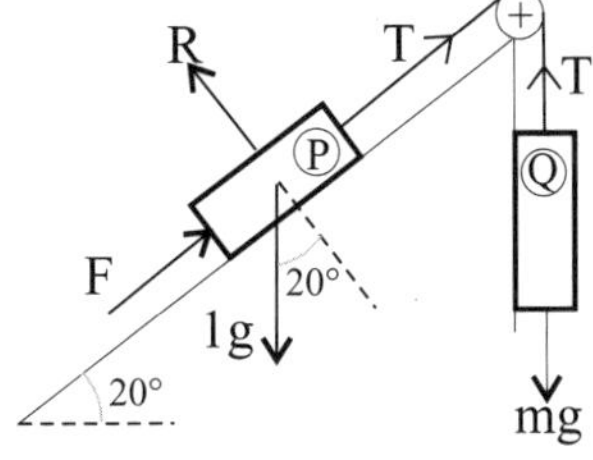

For Q: $F_{net} = ma$

Resolving vertically: $mg - T = 0$

$T = mg$ *[1 mark]*

For P: *Resolving in* ↖ *direction:*

$F_{net} = ma$

$R - 1g\cos 20° = 1 \times 0$

$R = g\cos 20°$ *[1 mark]*

Limiting friction: $F = \mu R$

$F = 0.1 \times g\cos 20°$ *[1 mark]*

Resolving in ↙ *direction:*

$1g\sin 20° - F - T = 1 \times 0$ *[1 mark]*

$1g\sin 20° - 0.1g\cos 20° - mg = 0$

$\sin 20° - 0.1\cos 20° = m$

$m = 0.248\ kg$ *[1 mark]*

b) *If Q = 1kg:*

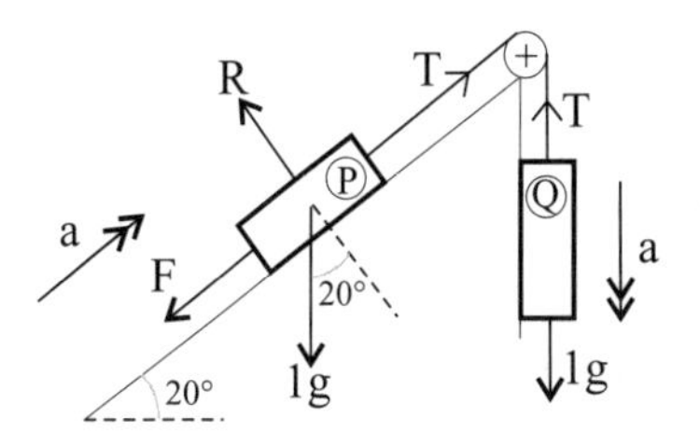

$F_{net} = ma$

For Q: resolving vertically: $1g - T = 1a$

$T = g - a$ ① *[1 mark]*

For P: resolving in ↖ *direction:* $R = g\cos 20°$

$F = \mu R = 0.1g\cos 20°$ *[1 mark]*

Resolving in ↗ *direction:* $T - 1g\sin 20° - F = 1a$

$T - 1g\sin 20° - 0.1g\cos 20° = 1a$ ② *[1 mark]*

Sub ① into ②:

$(g - a) - g\sin 20° - 0.1g\cos 20° = a$ *[1 mark]*

$g - g\sin 20° - 0.1g\cos 20° = 2a$

$5.527 = 2a$

$a = 2.76\ ms^{-2}$ *[1 mark]*

i.e. the masses move with an acceleration of $2.76\ ms^{-2}$

Page 35

1) $(5 \times 3) + (4 \times 1) = (5 \times 2) + (4 \times v)$

$19 = 10 + 4v$

$v = 2\tfrac{1}{4}\ ms^{-1}$ *to the right*

2) $(5 \times 3) + (4 \times 1) = 9v$

$19 = 9v$

$v = 2\tfrac{1}{9}\ ms^{-1}$ *to the right*

3) $(5 \times 3) + (4 \times -2) = (5 \times -v) + (4 \times 3)$

$7 = -5v + 12$

$5v = 5$

$v = 1\ ms^{-1}$ *to the left*

4) $(m \times 6) + (8 \times 2) = (m \times 2) + (8 \times 4)$

$6m + 16 = 2m + 32$

$4m = 16$

$m = 4\ kg$

5) ***Before*** ***After***

(0.8) →4 (1.2) →2 (0.8) →2.5 (1.2) →v

$(0.8 \times 4) + (1.2 \times 2) = (0.8 \times 2.5) + 1.2v$ *[1 mark]*

$3.2 + 2.4 = 2.0 + 1.2v$

$v = 3\ ms^{-1}$ *[1 mark]*

Before ***After***

(1.2) →3 4← (m) (1.2+m) 0

$(1.2 \times 3) + (m \times -4) = (1.2 + m) \times 0$ *[1 mark]*

$3.6 = 4m$

$m = 0.9kg$ *[1 mark]*

Answers

Page 37

1) *Impulse acts against motion, so* $I = -2$ Ns

$I = mv - mu$

$-2 = 0.3v - (0.3 \times 5)$

$v = -1\frac{2}{3}\ ms^{-1}$

2) *You need to find the particle's velocities just before and just after impact.*

Falling (down = +ve):

$u = 0$, $s = 2$, $a = 9.8$, $v = ?$

$v^2 = u^2 + 2as$

$v = \sqrt{2 \times 9.8 \times 2}$

$= 6.261\ ms^{-1}$

Rebound (this time, let up = +ve):

$v = 0$, $u = ?$, $a = -9.8$, $s = 1\frac{1}{3}$

$v^2 = u^2 + 2as$

$0 = u^2 + 2 \times -9.8 \times 1\frac{1}{3}$

$u = 5.112\ ms^{-1}$

Taking up = +ve:

Impulse $= mv - mu$

$= (0.45 \times 5.112) - (0.45 \times -6.261)$

$= 5.12$ Ns

3)a) Before / After

(4000) $\xrightarrow{2.5}$ (1000) 0 (5000) $\xrightarrow{v}$

$(4000 \times 2.5) + (1000 \times 0) = 5000v$ [1 mark]

$v = 2\ ms^{-1}$ [1 mark]

b) Impulse $= mv - mu$

$= (4000 \times 2) - (4000 \times 2.5)$ [1 mark]

$= -2000$ Ns [1 mark]

c) *Track horizontal; no resistance (e.g. friction) to motion; wagons can be modelled as particles — or any other valid assumption.*

[1 mark each for any two]

Section Five — Projectiles

Page 39

1) *Resolving horizontally:*

$u = 120;\ s = 60;\ a = 0;\ t = ?$

$s = ut + \frac{1}{2}at^2$

$60 = 120t + \frac{1}{2} \times 0 \times t^2$

$t = 0.5$ s

Resolving vertically: $u = 0;\ s = ?;\ a = 9.8;\ t = 0.5$

$s = ut + \frac{1}{2}at^2$

$= (0 \times 0.5) + (0.5 \times 9.8 \times 0.5^2)$

$= 1.23$ m (to 3 s.f.)

2) *Resolving horizontally:*

$u = 20\cos 30°;\ s = 30;\ a = 0;\ t = ?$

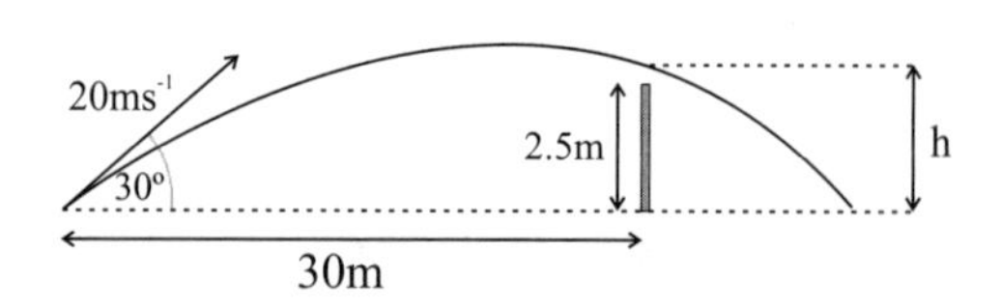

$s = ut + \frac{1}{2}at^2$ [1 mark]

$30 = (20\cos 30° \times t) + (\frac{1}{2} \times 0 \times t^2)$

$t = 1.732$ s [1 mark]

Resolving vertically:

$s = h;\ u = 20\sin 30°;\ t = 1.732;\ a = -9.8$

$s = ut + \frac{1}{2}at^2$ [1 mark]

$h = (20\sin 30° \times 1.732) + (\frac{1}{2} \times -9.8 \times 1.732^2)$

$= 2.62$ m (to 3 s.f.) [1 mark]

Therefore the ball goes over the crossbar. [1 mark]

Assumptions: ball is a point mass, no air resistance/wind. [1 mark]

Page 41

1) *First find time of flight to top:*

$v = 0;\ u = u\sin\alpha;\ a = -g,\ \alpha = ?$

$v = u + at$

$0 = u\sin\alpha - gt$

$t = \frac{u\sin\alpha}{g}$

(this is time taken to reach highest point — i.e. half way)

So $T = \frac{2u\sin\alpha}{g}$

Now plug in the values to get α.

$u = 22;\ g = 9.8;\ T = 4;\ \alpha = ?$

$4 = \frac{2 \times 22 \times \sin\alpha}{9.8}$

$\sin\alpha = 0.89$

$\alpha = 63.0°$ (to 3 s.f.)

2)a) $u_x = 50\cos 25° = 45.3\ ms^{-1}$ [1 mark]

$u_y = 50\sin 25° = 21.1\ ms^{-1}$ [1 mark]

b) *Resolving vertically:*

$u = 50\sin 25°;\ a = -9.8;\ t = 3;\ s = ?$

$s = ut + \frac{1}{2}at^2$ [1 mark]

$= (50\sin 25° \times 3) + (\frac{1}{2} \times -9.8 \times 9)$

$= 19.3$ m [1 mark]

Resolving horizontally:

$u = 50\cos 25°;\ a = 0;\ t = 3;\ s = ?$

$s = ut$ [1 mark]

$= 50\cos 25° \times 3$

$= 136$ m [1 mark]

So the target is 136 m away (horizontally) and is 19.3 m above the cannon.

Answers

c) $s = h;\ a = -9.8;\ u = 50\sin 25°;\ v = 0$

$v^2 = u^2 + 2as$ [1 mark]

$0 = (50\sin 25°)^2 + (2 \times -9.8 \times h)$

$$h = \frac{(50\sin 25°)^2}{2(9.8)}$$

$= 22.8\ m$ [1 mark]

d) *Resolving vertically:*

$u = 50\sin 25°;\ a = -9.8;\ t = 3;\ v = ?$

$v = u + at$

$v = 50\sin 25° - (9.8 \times 3)$

$= -8.269\ ms^{-1}$ [1 mark]

Resolving horizontally:

$u = 50\cos 25°;\ t = 3;\ v = ?;\ a = 0$

$v = u + at$

$= 50\cos 25° + (0 \times 3)$

$= 45.315$ [1 mark]

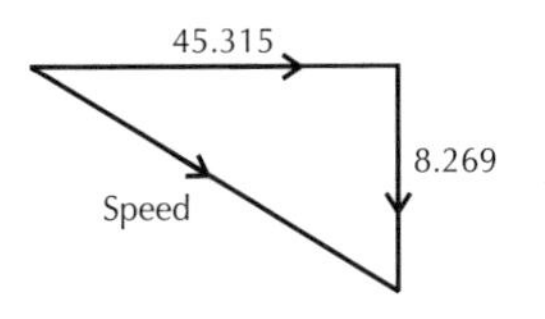

$\text{Speed} = \sqrt{45.315^2 + 8.269^2}$

$= 46.1\ ms^{-1}$ [1 mark]

Page 43

1)

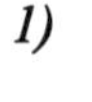

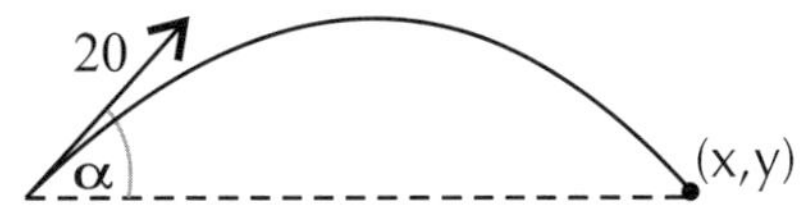

Use the equation $y = x\tan\alpha - \dfrac{gx^2}{2u^2\cos^2\alpha}$

(make sure you know the derivation of the equation)

Use identity $\dfrac{1}{\cos^2\alpha} = \sec^2\alpha = (1 + \tan^2\alpha)$:

$$y = x\tan\alpha - \frac{gx^2}{2u^2}\left(1 + \tan^2 a\right)$$

a) $(x, y) = (30, 0)$

$0 = 30\tan\alpha - \dfrac{10 \times 30^2}{2 \times 20^2}\left(1 + \tan^2 a\right)$

$0 = 120\tan\alpha - 45(1 + \tan^2\alpha)$

$3\tan^2\alpha - 8\tan\alpha + 3 = 0$ — *a quadratic in* $\tan\alpha$*, so use the quadratic formula:*

$$\tan\alpha = \frac{8 \pm \sqrt{64 - 4(9)}}{6}$$

$\alpha = 65.7°$ or $\alpha = 24.3°$

b) $(x, y) = (40, 0)$

$0 = 40\tan\alpha - \dfrac{10 \times 40^2}{2 \times 20^2}\left(1 + \tan^2 a\right)$

$0 = 40\tan\alpha - 20(1 + \tan^2\alpha)$

$\tan^2\alpha - 2\tan\alpha + 1 = 0$

$$\tan\alpha = \frac{2 \pm \sqrt{4 - 4}}{2}$$

$\alpha = 45°$

c) $(x, y) = (50, 0)$

$0 = 50\tan\alpha - \dfrac{10 \times 50^2}{2 \times 20^2}\left(1 + \tan^2 a\right)$

$0 = 50\tan\alpha - \dfrac{125}{4}(1 + \tan^2\alpha)$

$5\tan^2\alpha - 8\tan\alpha + 5 = 0$

$$\tan\alpha = \frac{8 \pm \sqrt{64 - 100}}{10}$$

$$= \frac{8 \pm \sqrt{-36}}{10}$$

So there's no solution — i.e. (50, 0) is out of range.

2)

Use the equation $y = x\tan\alpha - \dfrac{gx^2}{2u^2\cos^2\alpha}$

Sub in $\left(\dfrac{1}{\cos^2 a}\right) = \sec^2 a$ *and* $\sec^2 a = \left(1 + \tan^2 a\right)$:

So $y = x\tan\alpha - \dfrac{gx^2}{2u^2}(1 + \tan^2 a)$

$x = 12,\ y = 6,\ g = 10,\ u = 10$

Substituting these values:

$6 = 12\tan\alpha - \dfrac{10 \times 12^2}{2 \times 20^2}(1 + \tan^2 a)$

$6 = 12\tan\alpha - \dfrac{9}{5}(1 + \tan^2\alpha)$

$3\tan^2\alpha - 20\tan\alpha + 13 = 0$

$$\tan\alpha = \frac{20 \pm \sqrt{244}}{6}$$

$\alpha = 36.1°$ or $\alpha = 80.4°$

3)

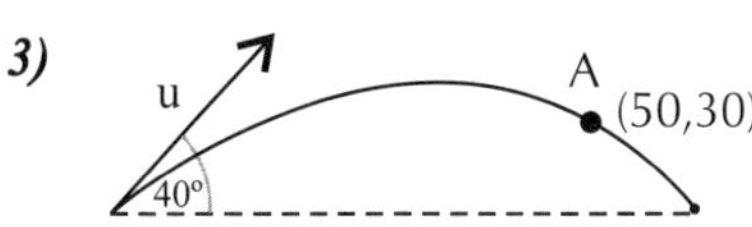

$y = x\tan\alpha - \dfrac{gx^2}{2u^2\cos^2\alpha}$ [2 marks for derivation of equation]

$30 = 50\tan 40° - \dfrac{10 \times 50^2}{2u^2 \times \cos^2 40°}$ [1 mark]

$$u^2 = \frac{25000}{2\cos^2 40°(50\tan 40° - 30)} = \frac{25000}{1.174 \times 11.955}$$

$u = 42.2\ ms^{-1}$ [1 mark]

Resolving horizontally:

$s = 50$

$u_x = 42.2 \times \cos 40°$

$s = ut$ so $50 = 42.2 \times \cos 40° \times t$ [1 mark]

$t = 1.55\ s$ (to 3 s.f.) [1 mark]

Index